Metrics
for the
Millions

By

Rufus P. Turner, Ph.D

HOWARD W. SAMS & CO., INC.
THE BOBBS-MERRILL CO., INC.
INDIANAPOLIS · KANSAS CITY · NEW YORK

Preface

The United States and Great Britain are now moving toward adoption of the metric system of weights and measures. Britain expects to be totally on the metric system by 1975; the U.S. will take longer. Prior to the switch from our present outdated system of feet, pounds, gallons, miles, etc., a great many Americans will need to learn the metric system.

Through comparison with the U.S. system, this book strives to explain the metric system in simplified terms—but, we hope, not oversimplified ones. We have in mind both the absolute newcomer who needs an introduction, and the old hand who wants a refresher.

The various classes of metric units (length, area, capacity, etc.) are covered in successive chapters which have parallel development. In each chapter, illustrative examples show how to convert from metric to U.S., and from U.S. to metric. A set of practice exercises with which the reader may test his skill is offered at the end of each chapter (the answers to all practice exercises are given in Appendix 1). At the end of the book, Appendix 2 gives an alphabetically arranged collection of conversion factors for the easy conversion of metric units to U.S., and of U.S. units to metric. This latter section will make the book a handy reference manual to serve the reader long after he or she has learned the basics of the metric system.

RUFUS P. TURNER

To MARY

On Our 45th Anniversary

Contents

CHAPTER 1

Introduction

If you have taken a science course—especially chemistry or physics—in high school or college, the metric system is no stranger to you. You measured your lengths in *meters* or *centimeters*, not feet or yards; weighed your materials in *grams*, not pounds or ounces; and portioned out your liquids in *liters* or *cubic centimeters*, not pints or quarts. In short, you regularly used the international system of units. But outside of the classroom and laboratory, you lived naturally with feet, pounds, quarts, bushels, and the other nonmetric units which are standard in the everyday life of the United States.[1]

Many people, however, know nothing of the metric system; and some others who did know but have not followed a scientific career since finishing school, have forgotten most of the details of the system. This chapter is addressed to both groups; its purpose is to introduce or to refresh, whichever is required.

---

[1] It is common to call these nonmetric units *British units* or *English units*. But since there is a difference in value between some of the nonmetric units used in England and those used in the United States, the term *U.S. units* is used throughout this book.

Since the United States now seems on the way to joining the majority of the countries of the world in adopting the metric system—in place of the cumbersome nonmetric system—for everyday use, many Americans will want to familiarize themselves quickly with the metric system.

1.1. What Is the Metric System?

The metric system is a decimal system of weights and measures, employing *grams, kilometers, liters,* and other such units. In many respects, it is much easier to handle than is the U.S. system of pounds, miles, gallons, and so on. The metric system is in widespread general use throughout the world, except notably in the United States where it is used in scientific measurements, but not for everyday purposes.

1.2. The Decimal System, Ancestor of the Metric System

Since the metric system is a decimal-based system, the Decimal System of Notation itself merits a brief review here. To begin with, the decimal system is a method of counting based on the number 10. Its name derives from the Latin word *decem* meaning "ten." It employs zero and nine digits (0, 1, 2, 3, 4, 5, 6, 7, 8, 9) from which all numbers—from exceedingly tiny to fantastically huge— can be formed. There are many other ways of counting, but for several centuries the decimal system has proved convenient and serviceable in a variety of ways throughout the world.

Scholarship and folklore both try to account for the origin of the decimal system. It is said, for example, that this system grew naturally out of man's endowment of ten fingers (see Fig. 1-1). Be that as it may, however, historians of mathematics tell us that the system originated in India, that it was revived in Europe via the Arabs, and that it was explained exhaustively in 1202 by Leonardo da Pisa (also called Leonardo Fiboracci or Pisano), an Italian geometrician.

Decimal fractions (such as 0.656) are now old hat to almost every schoolboy and are regularly used by a respectable number of adults who find them more congenial than regular fractions (such as 21/32). They were invented around 1585 by Simon Stevin (also known as Stevinus), a Dutch mathematician who died in the same year that the Pilgrims landed at Plymouth. This pioneer lobbied for adoption by his government of a decimal system of weights and measures. But the metric system was not adopted until some 200 years later.

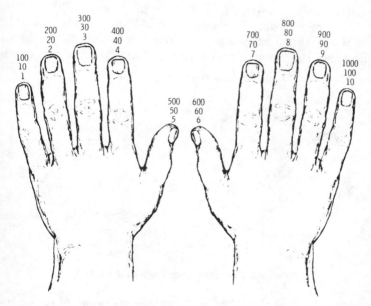

Fig. 1-1. The decimal system is a natural method.

Today, it is hard to imagine that the decimal system of counting was ever unknown to man. Its usefulness extends into many areas of public and private life. For example, when we talk about a tenth of some quantity, we are more likely to visualize the decimal fraction 0.1 than the common fraction 1/10. The U.S. money system is a decimal system.

1.3. Birth and Growth of the Metric System

The metric system as a government-prescribed decimal system of weights and measures first appeared in 18th-century France. There, in 1791, the National Assembly received and sometime later put into effect the provisions of a report from the French Academy standardizing the *meter* as the unit of length and the *gram* as the unit of mass and weight. ("Meter" is derived from the Greek word *metron*, meaning "a measure," and "gram" from the Latin word *gramma*, meaning "a small weight.")

The meter was eventually defined as one ten-millionth of the distance (at sea level) along a meridian of the earth between the equator and a pole (Fig. 1-2). And the gram

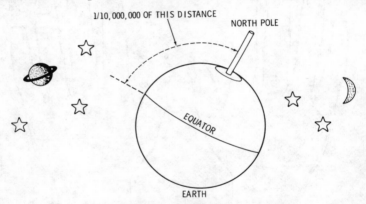

1/10,000,000 OF THIS DISTANCE

NORTH POLE

EQUATOR

EARTH

Fig. 1-2. First standardization of the meter.

was defined as the mass of 1 cubic centimeter (i.e., a cube whose sides are each 1/100 of a meter long) of pure water at the temperature of 4 degrees centigrade (the temperature at which water is most dense). See Fig. 1-3. Once these values were agreed upon, they could be multiplied or divided at will to describe larger and smaller quantities.

In 1793, the standard meter was physically actualized in this manner: On a bar of platinum-iridium alloy having an X-shaped cross section, two lines were inscribed 1 meter apart; this separation (when the temperature of

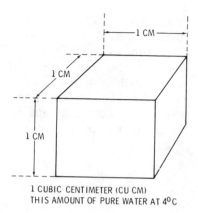

Fig. 1-3. First standardization
of the gram.

1 CUBIC CENTIMETER (CU CM)
THIS AMOUNT OF PURE WATER AT 4°C

the bar is zero degrees centigrade) being equal to one ten-millionth of the great-circle distance from the equator to a pole, as explained above. This distance was determined by means of a series of geodetic measurements by France and England that are too specialized for explanation here. The prototype standard meter is kept at the International Bureau of Weights and Measures at Sèvres, a suburb of Paris, France. Two other such bars were made and inscribed, to be kept elsewhere. Copies of the International Prototype Meter are owned by governments all over the world, including the United States. Ours, shown in Fig. 1-4, is kept at the National Bureau of Standards. In 1960, the meter was redefined more accurately and conveniently in terms of the wavelength of light in a vacuum, as being equal to 1,605,763.73 wavelengths of the orange-red light from the isotopic gas krypton 86. Today, highly sophisticated instruments are available for measuring the wavelength of this light emitted by a special electron tube containing the krypton isotope. This is a more satisfactory procedure than remeasuring a meridian of the earth and dividing it into ten million equal parts.

The unit of mass was next physically actualized in another way other than as the mass of 1 cubic centimeter of water. A block of the same platinum-iridium alloy from which the Protoype Meter was made, was cut and finished to certain prescribed dimensions to give it a mass of 1000 grams (i.e., 1 *kilogram*). This International Prototype Kilogram likewise is kept at the International Bureau of Weights and Measures at Sèvres, France; and copies of it are owned by governments all over the world, including the United States. Ours, like our copy of the International Prototype Meter, is kept at the National Bureau of Standards. Our copy of the Prototype Kilogram is shown in Fig. 1-5.

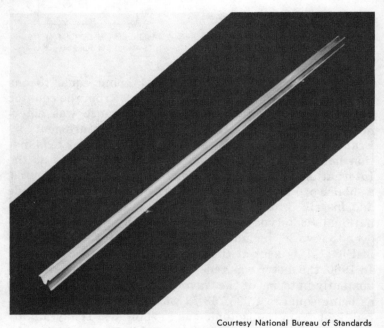

Courtesy National Bureau of Standards

Fig. 1-4. U.S. copy of the Prototype Meter on deposit at the National Bureau of Standards.

In 1866, the U.S. Congress authorized optional use of the metric system in this country. But up to the time of this writing, we have not officially embraced the metric system for general use, although we refer our inch, pound, and other nonmetric units to our copies of the Prototype Meter and Prototype Kilogram as standards.

The first step toward worldwide adoption of the metric system was taken in 1872 when the French government called an international conference to which 26 nations, including the United States, sent representatives. An international treaty termed the *Metric Convention* (also called the *Treaty of the Meter*) grew out of this conference and in 1875 was signed by 18 countries, among them the United States. Since that time, a number of General Conferences on Weights and Measures have followed.

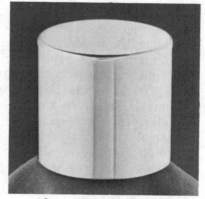

Fig. 1-5. Prototype Kilogram on deposit at the National Bureau of Standards.

Courtesy National Bureau of Standards

1.4. Aspects of the Metric System

Any measuring system must offer a wide range of use; the system must allow for the measurement of very large and very small quantities, as well as average-size ones. How this range is provided for and how the large and small units are named, determine to a large extent how clear or confusing the system may seem to those learning to use it.

In our present U.S. system, for example, we can start with inches for the measurement of length; and as we accumulate inches, we simplify things by calling every 12 inches a foot, every 36 inches (or 3 feet) a yard, and every 5280 feet (or 1760 yards) a mile (not forgetting, of course, that there are also some lesser-known units, such as the rod, furlong, and league). Now, in applying this method, we must not only remember the multipliers, which bear no simple relationship to each other, but also unrelated names (such as "inch" and "foot") which disappear in the process. These requirements do nothing to ease the job of learning the system or using it. In weight measure, we have a similar problem as we move from ounces to pounds to tons.

The metric system is far neater. In it, we never lose the name of the basic unit. We merely attach a prefix to the original name to indicate whether we are multiplying or dividing the basic unit and by how much. We need only remember the meaning of each prefix. For example, the basic unit of length is the meter. As we accumulate meters, we simplify matters in this way: 10 meters make 1 decameter, 100 meters make 1 hectometer, 1000 meters make 1 kilometer, and so on. Conversely, as we divide the meter, we follow a similar procedure: 1/10 meter is 1 decimeter, 1/100 meter is 1 centimeter, 1/1000 meter is 1 millimeter, and so on. Notice that the basic unit, *meter*, never leaves the stage. The very same prefixes are used with other units of measurement. Thus, 1 kilogram = 1000 grams, 1 milligram = 1/1000 gram, 1 hectoliter = 100 liters, and so on. Table 1-1 lists all of the prefixes. Table 1-2 lists metric units and their abbreviations, and Table 1-3 lists abbreviations and the corresponding metric units.

1.5. The International System of Units (SI)

The International System of Units (SI)[2] is the metric system of measurements which the 11th General Conference on Weights and Measures, with representatives from 36 nations meeting in Paris, established in 1960 under the Treaty of the Meter. In this system, there are six basic units: the *meter* (m) for length, the *kilogram* (kg) for mass, the *second* (s) for time, the *kelvin* (K) for tem-

---

[2]Here, *SI* stands for the French term "Système International (d'Unités)."

perature, the *ampere* (A) for electric current, and the *candela* (cd) for luminous intensity. All other units are obtained from these six, by derivation or combination.

Table 1-1. Metric Prefixes

Prefix	Symbol	Meaning
atto-	a	$1/10^{18}$*
centi-	c	$1/100$
deca- (also *deka*-)	dk	$\times 10$
deci-	d	$1/10$
femto-	f	$1/10^{15}$†
giga-	G	$\times 1,000,000,000$
hecto- (also *hekto*-)	h	$\times 100$
kilo-	k	$\times 1000$
mega-	M	$\times 1,000,000$
micro	μ	$1/1,000,000$
milli-	m	$1/1000$
myria-	my	$\times 10,000$
nano-	n	$1/1,000,000,000$
pico-	p	$1/10^{12}$‡
tera-	T	$\times 10^{12}$§

*1 quintillionth   †1 quadrillionth   ‡1 trillionth   §1 trillion

The meter is defined precisely as 1,650,763.73 wavelengths in vacuum "corresponding to the transition $2p_{10}$–$5d_5$ of the isotopic gas krypton 86" (see also the reference in Section 1.3 regarding the modern definition of the meter in terms of the wavelength of light). This definition of the meter, in addition to being more accurate than the old one, is based on a standard which is more universally available than are official copies of the meter bar. For ordinary practical purposes, however, the difference between the old and new value is very slight—4 ten-thousandths of 1 percent.

The kilogram is still defined as the mass of the International Prototype Kilogram preserved at Sèvres, France (see Section 1.3). But it should be noted that in the SI

Table 1-2. Metric Units & Abbreviations

Unit	Abbreviation	Unit	Abbreviation
are	a	hectometer	hm
centare	ca	kilogram	kg
centigram	cg	kiloliter	kl
centiliter	cl	kilometer	km
centimeter	cm	liter	l
cubic	cc (also cu cm	meter	m
centimeter	or cm³)	metric ton	MT (also t)
cubic meter	cu m (also m³)	milligram	mg
decagram	dkg	milliliter	ml
decaliter	dkl	millimeter	mm
decameter	dkm	myriameter	mym
decastere	dks	quintal	q
decigram	dg	square	sq cm
deciliter	dl	centimeter	(also cm²)
decimeter	dm	square	sq km
decistere	ds	kilometer	(also km²)
gram	g (also gm)	square meter	sq m
hectare	ha		(also m²)
hectogram	hg	stere	s
hectoliter	hl		

scheme, the kilogram is the unit of mass only, not of weight. The unit of weight and of force is the *newton* (N). (1 N = 1 kg · m/s².)

The second (formerly officially defined as 1/86,400 of the mean solar day) is defined as "the duration of 9,192,631,770 periods of the radiation corresponding to the transition between the two hyperfine levels of the ground state of cesium 133."

The kelvin is defined as "1/273.16 of the thermodynamic temperature of the triple point of water." Temperature in degrees Celsius (formerly called *centigrade*) is easily converted to temperature in kelvins by adding 273.16 to the Celsius figure.

Of particular interest to electronics people, the ampere is defined as the electric current which, flowing in two infinitely long parallel wires situated in a vacuum and separated by a distance of 1 meter, produces between the wires a force of 2×10^{-7} newton per meter of wire length.

The candela is defined as "the luminous intensity of 1/600,000 square meter of a perfect radiator at the temperature of freezing platinum."

1.6. U.S. Movement Toward Adoption of the Metric System

Although we subscribed to the Metric Convention, were active in the General Conferences, and preserved metric standards of weights and measures, the United States for more than 300 years has held to the relatively cumbersome, nonmetric, British system in the everyday life of the country (England herself is making the changeover). However, numerous campaigns have been waged for

Table 1-3. Metric Abbreviations & Units

Abbreviation	Unit	Abbreviation	Unit
a	are (also prefix atto-)	hl	hectoliter
		hm	hectometer
ca	centare	kg	kilogram
cc	cubic centimeter	kl	kiloliter
cg	centigram	km	kilometer
cl	centiliter	km²	square kilometer
cm	centimeter	l	liter
cm²	square centimeter	m	meter (also prefix milli-)
cm³	cubic centimeter		
cu cm	cubic centimeter	m²	square meter
cu m	cubic meter	m³	cubic meter
dg	decigram	mg	milligram
dkg	decagram	ml	milliliter
dkl	decaliter	mm	millimeter
dkm	decameter	MT	metric ton
dks	decastere	mym	myriameter
dl	deciliter	q	quintal
dm	decimeter	s	stere
ds	decistere	sq cm	square centimeter
g	gram	sq km	square kilometer
gm	gram	sq m	square meter
ha	hectare	t	metric ton
hg	hectogram		

adoption of the metric system, and the American public has never been completely ignorant of this system. For example, while we still buy our vegetables by the pound or peck, we are told the potency of our vitamin pills and the nicotine content of our cigarettes in *milligrams*, even if aspirin and saccharin tablets are rated in the nonmetric *grains*. A chic restaurant will suggest half a *liter* of wine with your dinner, but you buy a quart of whiskey or a gallon of gasoline. Many inexpensive rulers have long carried a *centimeter* scale opposite the inch scale. And we can buy 35-*millimeter* film for our camera.

In spite of this coexistence of the two systems in our society, metric terms seem difficult and "unnecessarily European" to some Americans. Eventually, however, we will ask for a liter of milk as naturally as we now request a quart. In the interim, it is interesting to note that a growing number of suppliers now label their products in metric, as well as nonmetric, units in anticipation of the eventual changeover. For example, some of the food products this author has noticed which have net weight printed in both grams and ounces on their containers include: cookies, crackers, corn starch, spices, and peanuts. Some tomato juice and other canned goods are labeled in milliliters in addition to the nonmetric fluid ounces. With practice, we shall find the 50 kilometers distance to the next town as familiar an expression as the equivalent 31 miles or so. And as one wag has suggested, a bathing beauty's measurements of 96.5–60.9–93.9 (centimeters, that is) will rest as lightly on our ears (and eyes!) as the present 38–24–37 (inches, that is).

Some American firms have for many years had to work from metric plans and to metric specifications, especially when manufacturing for the foreign market. In their shops, the U.S. and metric systems have lived side by side with no great friction. Some entrepreneurs in the export business likewise have handled both systems with virtuosity. To these groups, a complete metric changeover will cause little of the inconvenience anticipated by rank newcomers. A major factor to be dealt with, of course, is the inertia resulting from custom and habit. Thus, many of the persons who have been measuring by the U.S. system for a lifetime will not gracefully lay aside their inches and pounds; for them the two systems will always exist side by side, legislation notwithstanding. It is to be expected, however, that children coming to the subject of measurements for the first time will have less trouble than some of their elders embracing the metric system. The metric system is not difficult to learn (in many respects, it is easier than the U.S. system) and the children will not have to unlearn a nonmetric system as their elders will.

No one pretends that a complete switch to the metric system will not be costly. Innumerable machines, tools, and instruments will need to be altered; dials and scales changed; plans redrawn (or at least relabeled); specifications, books, pamphlets, deeds, mortgages, descriptions, etc. rewritten; signs repainted; laws rephrased; maps redrawn; and so on. Then, there is the task of re-educating a great mass of people. All of these grand-scale activities will be expensive and demanding, to be sure. In the long run, however, it may prove more expensive to remain out of step with the rest of the world in this matter.

At the time of this writing, a number of bills before the U.S. Congress concern adoption of the metric system; one of these shoots at full conversion by 1984. In August 1973, the California Division of Oil and Gas—a state government agency—switched to the metric system and is believed to be the first such agency in the United States to have done so.

CHAPTER 2

Units of Length

Americans touring continental Europe soon observe that road signs there give distances and speed limits in *kilometers*. This leads occasionally to the flippant comment that a kilometer "is a European mile;" and if you take the quip with any degree of literalness, you might find yourself driving the American car you took along (with its speedometer reading in miles per hour) much faster than you are allowed to in some zones. Actually, a kilometer is a little bit longer than half a U.S. mile. Similarly, such tourists have been known to call the meter a "French yard" (the meter is approximately 3⅜ inches longer than a U.S. yard). The trouble with these simplifications is that the tourist is apt to keep in mind the familiar terms "mile" and "yard" without bothering himself with the actual differences between these terms and the metric units. A few travelers, of course, make the necessary calculations, so that they know the number of units involved in any situation.

Length (i.e., linear measure) is a good starting point in the study of any system of weights and measures. It is the fundamental dimension in the metric system. This chapter is devoted to the units of length and their conversion.

What Is the Basic Unit of Length in the Metric System?

The *meter*. In fact, the meter is the basis of the metric system which takes its name, "metric," from the meter and derives other units of measurement from the meter, the fundamental unit of length and distance in this system. Section 1.3 in Chapter 1 explains how the length of the meter was originally standardized and how today it is referred to the wavelength of a certain color of light. The meter is a little longer than a U.S. yard, being equal to 39.37 inches or about 3.281 feet (see Fig. 2-1). One yard = 0.914 meter; 1 inch = 0.0254 meter.

The meter is not a complete stranger to Americans. For many years, the transmissions of radio stations here were identified by their wavelengths in meters. Indeed, the popular term *short wave* perpetuates that usage, denoting a signal of short wavelength. Also, we send our contenders into the Olympic games to run in races a certain number of meters long.

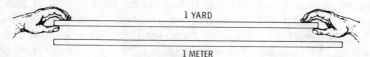

Fig. 2-1. Comparison of a yardstick and a meterstick.

Can This Unit Satisfy all Requirements?

It could—just as the foot might possibly satisfy all requirements in the U.S. system. For instance, we might say that the distance to the next town is 10,000 meters or that the length of a certain pencil is 14 hundredths of a meter. But there are better ways of specifying lengths or distances that are longer or shorter than 1 meter. This is explained in the next two sections.

How Is the Meter Multiplied for Longer Lengths?

We can express any longer distance or length as so many meters. But, just as when feet are used to express a long distance, we can run into cumbersome figures consisting of thousands of meters in some cases. Instead, we use decimal groupings of meters, naming each group as a new unit of measurement containing that number of meters. Thus, 10 meters (m) = 1 decameter (dkm), 100 meters = 1 hectometer (hm), 1000 meters = 1 kilometer (km), 10,000 meters = 1 myriameter (mym), and 1,000,000 meters = 1 megameter (Mm). See Table 1-1 in Chapter 1 for a list of prefixes attached to the word *meter*.

The kilometer is a convenient unit for the measurement of geographical distances. The road distance between New York and Chicago, for example, would be expressed in the metric system as 1310 kilometers (814 miles). One kilometer = 0.621 mile, or 3280.8 feet (see Fig. 2-2). The kilometer is useful also for expressing large values of height and depth. Thus, Mount Whitney, the highest point in the continental United States outside of Alaska, is 4.42 km (14,495 ft) high; and the lowest ocean depth on earth, the Mindanao Deep in the Pacific Ocean, is 10.79 km (35,400 ft) deep.

Table 2-1. Common Units of Length

Unit	Abbreviation	Meters
Megameter	Mm	1,000,000 meters
Myriameter	mym	10,000 meters
Kilometer	km	1000 meters
Hectometer	hm	100 meters
Decameter	dkm	10 meters
Meter	m	1 meter
Decimeter	dm	1/10 meter
Centimeter	cm	1/100 meter
Millimeter	mm	1/1000 meter
Micrometer	μm	1/1,000,000 meter

How Is the Meter Subdivided for Shorter Lengths?

Just as the meter is multiplied, as explained in the preceding section, it can also be subdivided into decimal groupings, each group being named as a new unit of measurement corresponding to that decimal subdivision. Thus, 1/10 meter (m) = 1 decimeter (dm), 1/100 meter (m) = 1 centimeter (cm)—see Figs. 2-3 and 2-4, 1/1000

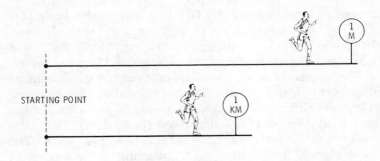

Fig. 2-2. Comparison of a mile and a kilometer.

meter (m) = 1 millimeter (mm), and 1/1,000,000 meter (m) = 1 micrometer (μm) also called 1 micron (μ). See Table 1-1 in Chapter 1 for a list of the prefixes that are attached to the word *meter*.

1 INCH DIVIDED INTO SIXTEENTHS

Fig. 2-3. Comparison of an inch and a centimeter.

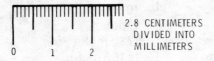

2.8 CENTIMETERS DIVIDED INTO MILLIMETERS

1 INCH = 2.54 CENTIMETERS

The centimeter is a convenient unit for the measurement of common lengths, being used somewhat in the manner that the inch is used in the U.S. system. One centimeter = 0.3937 inch or 0.0328 foot. One inch = 2.54 centimeters. A foot ruler is 30.48 cm long.

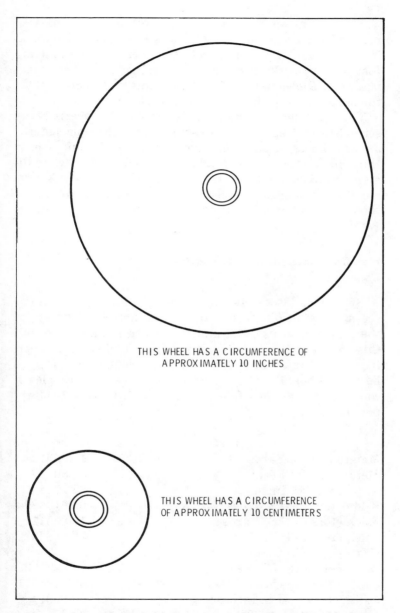

THIS WHEEL HAS A CIRCUMFERENCE OF
APPROXIMATELY 10 INCHES

THIS WHEEL HAS A CIRCUMFERENCE
OF APPROXIMATELY 10 CENTIMETERS

Fig. 2-4. Comparison of circular measure in centimeters and inches.

The millimeter is a convenient unit for the measurement of short lengths and thicknesses. A machinist working in the metric system checks thicknesses in millimeters just as one working in the U.S. system measures thicknesses in thousandths of an inch. One millimeter = 39.4 thousandths of an inch. One inch = 25.4 millimeters.

The micrometer (micron) is employed, especially by scientists and engineers, to express microscopic lengths and thicknesses, as when they state the thickness of a thin-film plating to be 1 micron (equal to 1/1000 millimeter or 3.94 hundred-thousandths of an inch). X-rays have a wavelength between 1 millionth of a micron and 1 thousandth of a micron, depending upon their hardness (ability to penetrate).

How Can Metric Units of Length Be Converted to U.S. Units?

By means of fairly simple multiplications. In some cases, the multipliers contain several figures. The more of the figures you use, the more accurate the answers, but this makes a longhand multiplication tedious. However, if you use one of the small electronic calculators that have become so popular, the operation will be easy. Examples of common conversions follow. Others will be found in Appendix 2.

To convert	Multiply by
Meters to U.S. Yards:	1.093611
Meters to U.S. Feet:	3.280833
Meters to U.S. Inches:	39.37
Kilometers to U.S. Miles:	0.6213699
Kilometers to U.S. Yards:	1093.611
Kilometers to U.S. Feet:	3280.833
Centimeters to U.S. Inches:	0.3937
Centimeters to U.S. Feet:	0.03280833
Centimeters to U.S. Yards:	0.01093611
Millimeters to U.S. Inches:	0.03937
Millimeters to U.S. Thousandths of Inch:	39.37

Illustrative Example: How many U.S. yards in 3.75 meters of cloth? Yards = meters × 1.093611. So, 3.75 × 1.093611 = 4.10 yards = 4 yd, 3.6 in.

Illustrative Example: How many U.S. miles in 150 kilometers? Miles = kilometers × 0.6213699. So, 150 × 0.6213699 = 93.2 miles.

Illustrative Example: How many thousandths of an inch in ½ millimeter? Thousandths of an inch = millimeters × 39.37. So, 0.5 × 39.37 = 19.7 thousandths.

How Can U.S. Units of Length Be Converted to Metric Units?

Again, by means of simple multiplications. As in the preceding case, however, some of the multipliers contain several figures, which can make a longhand multiplication tedious. But if you use a small electronic calculator, this is no problem. Examples of common conversions follow. Others may be found in Appendix 2.

To convert	*Multiply by*
U.S. Miles to Kilometers:	1.6093472
U.S. Miles to Meters:	1609.3472
U.S. Yards to Kilometers:	0.0009144
U.S. Yards to Meters:	0.9144018
U.S. Yards to Centimeters	91.44108
U.S. Feet to Kilometers:	0.0003048
U.S. Feet to Meters:	0.3048
U.S. Feet to Centimeters:	30.48
U.S. Inches to Meters:	0.0254
U.S. Inches to Centimeters:	2.54
U.S. Inches to Millimeters:	25.4
U.S. Thousandths of Inch to Millimeters:	0.0254

Illustrative Example: How many meters in an 18-yard typewriter ribbon? Meters = yards × 0.9144018. So, 18 × 0.9144018 = 16.46 meters.

Illustrative Example: How many kilometers in 47 U.S. miles? Kilometers = miles × 1.6093472. So, 47 × 1.6093472 = 75.64 kilometers.

Illustrative Example: How many millimeters in 26 thousandths of an inch (0.026 inch)? Millimeters = thousandths of an inch × 0.0254. So, 26 × 0.0254 = 0.66 millimeter.

PRACTICE EXERCISES

1. Convert 5½ meters to U.S. yards.
2. Convert 75 kilometers to U.S. miles.
3. Convert 12.3 centimeters to U.S. inches.
4. Convert 110 millimeters to U.S. inches.
5. Convert 30 meters to U.S. feet.
6. Convert 900 U.S. miles to kilometers.
7. Convert 1 U.S. rod to meters.
8. Convert 5′6″ (U.S.) to meters.
9. Convert 12 U.S. inches to centimeters.
10. Convert ⅛ U.S. inch to millimeters.

CHAPTER 3

Units of Area

Common area or space measurements in the U.S. system extend from the *square inch* at the low end of the spectrum to the *square mile* at the high end. Similarly, area measurements in the metric system extend from the square centimeter to the square kilometer. In the U.S. system, we have no unit of area smaller than the square inch and must refer to smaller areas as so many decimal (or fractional) parts of a square inch; thus, "0.25 sq in" or "¼ sq in." In the metric system, however, units of area smaller than the square centimeter include the square millimeter· and the square micron (square micrometer). Also, the user of the metric system can refer to a small area as so many decimal parts of a square millimeter; thus, "0.1 sq mm."

Area is the next logical step after length in our study of the metric system, since area is the product of length and width (actually the product of two lengths). This chapter is devoted to the units of area and their conversion.

Which Is the Basic Unit of Area in the Metric System?

The *square meter* (also called the *centare*). This is equal in its simplest form to a square with both sides 1-meter long. But 1 square meter of area can be enclosed by any number of combinations of length and width; e.g., 2-m long and ½-m wide, 4-m long and ¼-m wide, and so on.

Which U.S. Unit Does the Square Meter Most Resemble?

Perhaps the *square yard*. Actually, the square meter is about 1.2 times larger than the square yard, and the

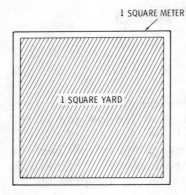

1 SQUARE METER

1 SQUARE YARD

Fig. 3-1. Comparison of the square meter and the square yard.

square yard is equal to approximately 84 percent of the square meter. See Fig. 3-1.

Can the Basic Metric Unit Satisfy all Requirements?

It could—just as the square inch might possibly satisfy all requirements of area measurement in the U.S. system—but for very large areas and very small ones, the numbers would grow unmanageable. A square mile in the U.S. system, for example, contains more than 4 billion square inches; and a square kilometer in the metric system is 1,000,000 square meters. There are better ways of expressing larger and smaller areas, and this is explained in the next two sections.

How Is the Square Meter Multiplied for Larger Areas?

While the square meter could be used to express all areas however large, our figures soon would grow into millions of square meters and therefore be cumbersome to handle. The same inconvenience arises in the U.S. system if we try to express huge areas in square inches, square feet, or even square yards (we have, instead, acres and square miles). The uniform decimal basis of the metric system permits us to move simply and easily to larger groupings of square meters, naming each group as a new unit of measurement containing that number of square meters. Thus, 100 square meters (sq m) = 1 square decameter (sq dkm), i.e., 10 meters on each side of a square (this area also has another name: 1 *are*); 10,000 square meters (sq m) = 1 hectare (ha) or 100 ares; 1,000,000 square meters (sq m) = 1 square kilometer (sq km) or 100 hectares. It should be noted that as each side of an enclosed square is increased, the area itself becomes multiplied by the square of the amount that the sides are increased. Thus, when we move from a square having 1 meter on each side to one having 10 meters on each side, the original area is multiplied 100 times. See Table 1-1 in Chapter 1 for a list of the prefixes that are attached to the words *meter* and *are*.

The square kilometer is a convenient unit for expressing large areas. Fig. 3-2 shows how the square kilometer compares with the U.S. square mile.

How Is the Square Meter Subdivided for Smaller Areas?

The square meter is subdivided decimally into smaller units of area measurement in much the same way that it is multiplied for larger areas. Each smaller decimal grouping is named as a new unit of measurement corresponding to that decimal subdivision. Thus, 1 square centimeter (i.e., a square having 1-cm sides) =0.0001 square meter, 1 square millimeter (i.e., a square having 1-mm sides) = 0.000001 square meter, and 1 square micron (i.e., a square

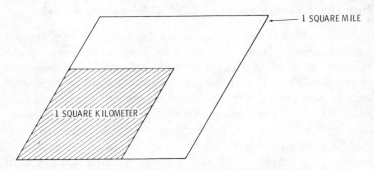

Fig. 3-2. A square kilometer compared with a U.S. square mile.

having 1/1,000,000 of a meter on each side) = 1/1,000,000,000,000 square meter. Fig. 3-3 shows the square centimeter compared with the U.S. square inch.

How Can Metric Units of Area Be Converted to U.S. Units?

By means of fairly simple multiplications. As in the conversion of lengths in Chapter 2, some of the multipliers contain several figures. The more of the figures you

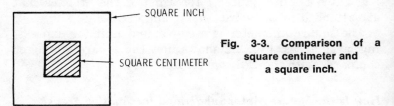

Fig. 3-3. Comparison of a square centimeter and a square inch.

use, the more accurate the answer. However, this makes longhand multiplication tedious. The operation may be eased by using one of the increasingly popular, small, electronic calculators. Examples of common conversions follow. Others may be found in Appendix 2.

Table 3-1. Common Units of Area

Unit	Abbreviation	Square Meters
Square kilometer	sq km	1,000,000 square meters
Hectare	ha	10,000 square meters
Are	a	100 square meters
Centare	ca	1 square meter
Square centimeter	sq cm	1/10,000 square meter
Square millimeter	sq mm	1/1,000,000 square meter

To convert	*Multiply by*
Square Centimeters to U.S. Square Inches:	0.15499
Square Millimeters to U.S. Square Inches:	0.0015499
Square Meters to U.S. Square Yards:	1.195985
Square Meters to U.S. Square Feet:	10.76387
Square Decameters to U.S. Square Yards:	119.5985
Square Decameters to U.S. Square Feet:	1076.387
Square Kilometers to U.S. Square Miles:	0.38610
Square Kilometers to U.S. Square Feet:	10, 763, 870
Square Kilometers to U.S. Acres:	247.1044
Ares to U.S. Acres:	0.02471044
Ares to U.S. Square Miles:	0.00003861
Hectares to U.S. Acres:	2.471044
Hectares to U.S. Square Miles:	0.0038610

Illustrative Example: How many U.S. square miles are in 100 ares (1 hectare)? Ares × 0.00003861 = square miles. So, 100 × 0.00003861 = 0.00386 square mile.

Illustrative Example: A certain parcel of land contains 0.125 square kilometers. What is this in U.S. acres? Square kilometers × 247.1044 = acres. So, 0.125 × 247.1044 = 30.89 acres.

Illustrative Example: A carpet has an area of 10.03 square meters. What is this in U.S. square feet?

Square meters × 10.76387 = square feet. So, 10.03 × 10.76387 = 107.96 square feet.

Illustrative Example: The size of a sheet of photographic film is 9 × 12 cm (an area of 108 square centimeters). What is the area of this film in U.S. square inches? Square centimeters × 0.15499 = square inches. So, 108 × 0.15499 = 16.74 square inches.

How Can U.S. Units of Area Be Converted to Metric Units?

Again, by means of fairly simple multiplications. As in the preceding section, some of the multipliers contain several significant figures, which might make longhand multiplication tedious. However, this will cause no difficulty if a simple electronic calculator is used. The more figures you employ, the more accurate the answer will be. Examples of common U.S.-to-metric conversions follow. Others may be found in Appendix 2.

To convert	*Multiply by*
U.S. Square Miles to Square Kilometers:	2.590
U.S. Square Miles to Hectares:	259.0
U.S. Square Miles to Ares:	25,900
U.S. Square Miles to Square Meters:	2,590,000
U.S. Acres to Square Kilometers:	0.00404687
U.S. Acres to Ares:	40.4687
U.S. Acres to Hectares:	0.404687
U.S. Square Yards to Square Meters:	0.83613
U.S. Square Yards to Square Decameters:	0.00836
U.S. Square Feet to Square Kilometers:	0.00000009
U.S. Square Feet to Square Meters:	0.092903
U.S. Square Inches to Square Centimeters:	6.45163
U.S. Square Inches to Square Millimeters:	645.163

Illustrative Example: The city of Los Angeles has an area of 442 square miles. What is the size of this city in square kilometers. Square miles × 2.590 = square kilometers. So, 442 × 2.590 = 1144.78 square kilometers.

Illustrative Example: A tract of land proposed for a real estate development contains 30 acres. What is the size of this tract in square kilometers? Acres × 0.00404687 = square kilometers. So, 30 × 0.00404687 = 0.1214 square kilometer.

Illustrative Example: A certain house has 1672 square feet of floor space. What is this in square meters? Square feet × 0.092903 = square meters. So, 1672 × 0.092903 = 155.33 square meters.

Illustrative Example: A standard sheet of typewriter paper (8½" × 11") has an area of 93.5 square inches. What is this in square centimeters? Square inches × 6.45163 = square centimeters. So, 93.5 × 6.45163 = 603.23 square centimeters.

PRACTICE EXERCISES

1. Convert 10,000 square kilometers to U.S. square miles.
2. Convert 1 square kilometer to U.S. square feet.
3. Convert 9 square meters to U.S. square feet.
4. Convert 1000 ares to U.S. acres.
5. Convert 50 square centimeters to U.S. square inches.
6. Convert 110 U.S. square miles to square kilometers.
7. Convert 30 U.S. acres to hectares.
8. Convert 33¼ U.S. square yards to square meters.
9. Convert 144 U.S. square feet to square meters.
10. Convert 50 U.S. square inches to square centimeters.

CHAPTER 4

Units of Volume

In the U.S. system, basic units of volume are the cubic inch and cubic foot. Familiar workaday units are the pint, quart, gallon, and barrel. Units having the same name—such as the ounce and pint—are often used in both liquid measure and dry measure, although they have different values. Thus, a liquid pint contains 28.9 cubic inches, whereas a dry pint contains 33.60 cubic inches. Dry and liquid U.S. units are therefore treated separately in this chapter.

The historical basic unit of volume in the metric system is the *liter,* and in the International System of Units (see Section 1.5 in Chapter 1) it is the *cubic meter.* Both the liter and the cubic meter may be multiplied or divided in the usual decimal manner to obtain larger or smaller units respectively; thus, deciliters, milliliters, cubic centimeters, and so on.

Volume is the next logical step after area in our study of the metric system, as volume is the product of length × width × height, which is to say the product of area and height. This chapter is devoted to the units of volume and their conversion.

Which Is the Basic Unit of Volume in the Metric System?

The *liter* is the historic one. This was originally defined as the volume occupied by 1 kilogram of water at a temperature of 4°C and an atmospheric pressure of 760 millimeters of mercury. In the International System of Units (SI), the *cubic meter* (also called a *stere,* abbreviated *s*) is the basic unit.

The liter is somewhat less than a U.S. dry quart and slightly more than a U.S. liquid quart. This relationship is shown in Fig. 4-1. (The liter is 90.81% of a dry quart, and the liquid quart is 94.63% of a liter.) One liter is equal to 0.001000027 cubic meter. The cubic meter in its simplest form is the volume of a cube that is 1 meter long, 1 meter wide, and 1 meter high (see Fig. 4-2).

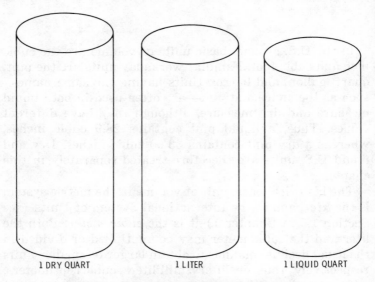

1 DRY QUART   1 LITER   1 LIQUID QUART

Fig. 4-1. Comparison of liter and U.S. quart.

Which U.S. Units Do These Units Most Resemble?

The liter closely resembles the U.S. quart (see the preceding section and Figure 4-1). The cubic meter closely re-

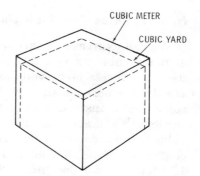

CUBIC METER

CUBIC YARD

Fig. 4-2. Comparison of cubic meter and cubic yard.

sembles the cubic yard; however, the cubic meter is approximately 1.31 times as large as the cubic yard. The comparison between the cubic centimeter and the cubic inch is shown in Fig. 4-3.

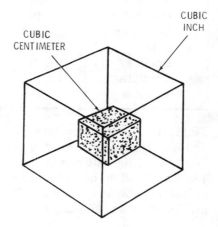

CUBIC INCH

CUBIC CENTIMETER

Fig. 4-3. Comparison of cubic inch and cubic centimeter.

Can the Basic Units Satisfy All Requirements?

The liter and the cubic meter could, just as the cubic inch might possibly satisfy all volume requirements in the U.S. system. A better method, however, is to use the standard metric method of multiplying or dividing the basic units decimally, as explained in the next two sections.

How Are These Units Multiplied for Larger Volumes?

We can express any larger volume as so many liters or so many cubic meters. But, just as when cubic inches are used to express large volumes, we can run into some unwieldy numbers consisting of millions of cubic meters in some cases. Instead, in the manner explained elsewhere for length and area, we use decimal groupings of liters or cubic meters, naming each group as a new unit of measure containing that number of cubic meters or liters. Thus, 10 liters (l) = 1 decaliter (dkl), 100 liters (l) = 1 hectoliter (hl), 1000 liters (l) = 1 kiloliter (kl), and so on.

Table 4-1. Common Units of Capacity

Unit	Abbreviation	Liters
Kiloliter	kl	1000 liters
Hectoliter	hl	100 liters
Decaliter	dkl	10 liters
Liter	l	1 liter
Deciliter	dl	1/10 liter
Centiliter	cl	1/100 liter
Milliliter	ml	1/1000 liter

Similarly, 10 cubic meters (cu m) = 1 cubic decameter (cu dkm), 100 cubic meters (cu m) = 1 cubic hectometer (cu hm), 1000 cubic meters (cu m) = 1 cubic kilometer (cu km), and so on. The cubic decameter is also called a *decastere*, abbreviated *dks*. The common units of volume are listed in Table 4-2.

Table 4-2. Common Units of Volume

Unit	Abbreviation	Cubic Meters
Decastere	dks	10 cubic meters
Stere	s	1 cubic meter
Decistere	ds	1/10 cubic meter
cubic centimeter	cu cm or cc	1/1,000,000 cubic meter

How Are These Units Subdivided for Smaller Volumes?

Just as the liter and cubic meter can be multiplied as explained in the preceding section, they can also be subdivided into decimal groupings, each group being named as a new unit of measure corresponding to that decimal subdivision. Thus, 1/10 liter (l) = 1 deciliter (dl), 1/100 liter (l) = 1 centiliter (cl), 1/1000 liter (l) = 1 milliliter (ml). Similarly, 1/10 cubic meter (cu m) = 1 cubic decimeter (cu dm), 1/100 cubic meter (cu m) = 1 cubic centimeter (cu cm or cc), 1/1000 cubic meter (cu m) = 1 cubic millimeter (cu mm). The cubic decimeter is also called a *decistere,* abbreviated *ds.*

The cubic centimeter (cu cm) and the milliliter (ml) are convenient units for a great many measurements. These two were intended to be equal, but close measurements showed that the liter (originally defined as the volume filled by 1 kg of water) was equal not to a cubic decimeter as intended, but to 1.000027 cubic decimeter. Hence, 1 milliliter = 1.000027 cubic centimeter, and 1 cubic centimeter = 0.999973 milliliter.

How Can Metric Units of Volume (Capacity) Be Converted to U.S. Units?

By means of fairly simple multiplications. In some cases, the multipliers, given below, contain several significant figures. The more of the figures you use, the more accurate the answers will be, but this makes a longhand multiplication tedious. However, if you use one of the small electronic calculators that have become so popular, the operation will be easy. Examples of common conversions follow. Other conversions from metric units of volume to U.S. units will be found in Appendix 2.

To convert	Multiply by
Cubic Meters to U.S. Cubic Yards:	1.3079
Cubic Meters to U.S. Cubic Feet:	35.314
Cubic Meters to U.S. Cubic Inches:	61,023

Cubic Meters to U.S. Bushels:	28.377
Cubic Meters to U.S. Barrels (Dry):	8.649
Cubic Meters to U.S. Barrels (Liq.):	8.3865
Cubic Meters to U.S. Gallons (Liq.):	264.173
Cubic Centimeters to U.S. Pints (Liq.):	0.00211
Cubic Centimeters to U.S. Pints (Dry):	0.001816
Cubic Centimeters to U.S. Quarts (Liq.):	0.001056
Cubic Centimeters to U.S. Quarts (Dry):	0.000908
Cubic Centimeters to U.S. Gallons (Liq.):	0.000264
Cubic Centimeters to U.S. Gallons (Dry):	0.000227
Cubic Centimeters to U.S. Fluid Ounces:	0.033808
Liters to U.S. Pints (Liq.):	2.1134
Liters to U.S. Pints (Dry):	1.8162
Liters to U.S. Quarts (Liq.):	1.0567
Liters to U.S. Quarts (Dry):	0.9081
Liters to U.S. Gallons (Liq.):	0.2641
Liters to U.S. Gallons (Dry):	0.22703
Milliliters to U.S. Fluid Ounces:	0.03381
Milliliters to U.S. Pints (Liq.):	0.00211
Milliliters to U.S. Quarts (Liq.):	0.00106

Illustrative Example: How many U.S. cubic yards in 1500 cubic meters? Cubic yards = cubic meters × 1.3079. Therefore, 1500 × 1.3079 = 1961.85 cubic yards.

Illustrative Example: How many U.S. liquid pints in 700 cubic centimeters? Liquid pints = cubic centimeters × 0.00211. Therefore, 700 × 0.00211 = 1.477 pint.

Illustrative Example: How many U.S. liquid gallons in 1.89 liter? Liquid gallons = liters × 0.2641. So, 1.89 × 0.2641 = 0.499 gallon (approximately ½ gallon).

Illustrative Example: How many U.S. fluid ounces in 36 milliliters? Fluid ounces = milliliters × 0.03381. So, 36 × 0.03381 = 1.217 fluid ounces.

How Can U.S. Units of Volume (Capacity) Be Converted to Metric Units?

Again, by means of simple multiplications. As in the preceding case, some of the multipliers contain several figures which can make a longhand multiplication tedious. If you use a small electronic calculator to perform the multiplications, this is not a problem. Examples of common conversions follow. Other conversions from U.S. units of volume to metric units may be found in Appendix 2.

To convert	Multiply by
U.S. Cubic Yards to Cubic Meters:	0.7645
U.S. Cubic Feet to Cubic Meters:	0.02832
U.S. Cubic Inches to Cubic Meters:	0.0000164
U.S. Bushels to Cubic Meters:	0.03524
U.S. Barrels (Dry) to Cubic Meters:	0.1156
U.S. Barrels (Liq.) to Cubic Meters:	0.1192
U.S. Pints (Liq.) to Cubic Centimeters:	473.18
U.S. Pints (Dry) to Cubic Centimeters:	550.61
U.S. Quarts (Liq.) to Cubic Centimeters:	946.36
U.S. Quarts (Dry) to Cubic Centimeters:	1101.23
U.S. Gallons (Liq.) to Cubic Centimeters:	3785.43
U.S. Gallons (Dry) to Cubic Centimeters:	4404.9
U.S. Fluid Ounces to Cubic Centimeters:	29.574
U.S. Pints (Liq.) to Liters:	0.47316
U.S. Pints (Dry) to Liters:	0.5506
U.S. Quarts (Liq.) to Liters:	0.9463
U.S. Quarts (Dry) to Liters:	1.1012
U.S. Gallons (Liq.) to Liters:	3.7853
U.S. Gallons (Dry) to Liters:	4.0476
U.S. Fluid Ounces to Milliliters:	29.573
U.S. Pints (Liq.) to Milliliters:	473.167
U.S. Quarts (Liq.) to Milliliters:	946.33

Illustrative Example: How many cubic meters in 2¼ U.S. bushels? Cubic meters = bushels × 0.03524. So, 2.25 × 0.03524 = 0.07929 cubic meter (approximately 1/13 cubic meter).

Illustrative Example: How many cubic centimeters in ¼ U.S. liquid pint? Cubic centimeters = liquid pints × 473.18. So, 0.25 × 473.18 = 118.29 cubic centimeters.

Illustrative Example: How many liters in 1½ U.S. liquid gallons? Liters = liquid gallons × 3.7853. So, 1.5 × 3.7853 = 5.678 liters.

Illustrative Example: A large can contains 16 fluid ounces of beer. What is this content in milliliters? Milliliters = fluid ounces × 29.573. So, 16 × 29.573 = 473.17 milliliters.

PRACTICE EXERCISES

1. Convert 55 cubic meters to U.S. cubic yards.
2. Convert 1250 cubic meters to U.S. cubic feet.
3. Convert 2 cubic meters to U.S. liquid barrels.
4. Convert 1100 cubic centimeters to U.S. liquid pints.
5. Convert 3300 cubic centimeters to U.S. liquid gallons.
6. Convert 150 cubic centimeters to U.S. fluid ounces.
7. Convert 15 liters to U.S. liquid quarts.
8. Convert ½ liter to U.S. liquid pints.
9. Convert 130 liters to U.S. liquid gallons.
10. Convert 75 milliliters to U.S. fluid ounces.
11. Convert 750 U.S. cubic yards to cubic meters.
12. Convert 1180 U.S. bushels to cubic meters.
13. Convert 2 U.S. dry barrels to cubic meters.
14. Convert 1½ U.S. liquid pints to cubic centimeters.
15. Convert 1/5 U.S. liquid gallon to cubic centimeters.
16. Convert 20 U.S. fluid ounces to cubic centimeters.
17. Convert ½ U.S. liquid pint to liters.
18. Convert 5 U.S. liquid gallons to liters.
19. Convert ½ U.S. fluid ounce to milliliters.
20. Convert ¼ U.S. liquid pint to milliliters.

CHAPTER 5

Units of Mass and Weight

Like the traveler at the beginning of Chapter 2 who encounters the kilometer for the first time on European highways and mistakenly thinks it is European for "mile," our tourist who checks his weight will find the scale reading in *kilograms*. But if he accepts with any degree of literalness the facetious quip, sometimes heard, that the kilogram is European for "pound," and if he is a weight watcher, he is in for a shock, for the kilogram is almost 2¼ U.S. pounds.

In metric indications of mass and weight, the neat decimal multiples and submultiples of the gram replace the less congenial ounces, pounds, drams, scruples, and grains of the U.S. system. The units of mass and weight are parallel to those of length, area, and volume, already studied in the preceding chapters.

This chapter is devoted to the units of mass and weight and to their conversions.

Which Is the Basic Unit of Mass in the Metric System?

The *gram*. All other units of mass and weight in the metric system are multiples or submultiples of the gram.

The international standard, however, is the *kilogram*, which is equal to 1000 grams. Section 1.3 in Chapter 1 explains how the value of the gram was arrived at and how the International Prototype Kilogram was fabricated.

In the International System of Units (SI), discussed in Section 1.5 of Chapter 1, the kilogram is the unit of mass only. A separate, derived unit of weight is the *newton* ($= \text{kg} \cdot \text{m/s}^2$).

How Do the Gram and Kilogram Compare With U.S. Units?

The gram is considerably smaller than a U.S. avoirdupois ounce—about 1/28 of an ounce (there are approximately 454 grams in 1 pound). There are 0.4536 kilograms in a U.S. avoirdupois pound, which makes the kilogram slightly more than 2.2 pounds. A 5-pound bag of sugar would be labeled 2268 grams. A 150-pound man weighs 68.04 kilograms.

Can the Basic Metric Unit Satisfy All Requirements?

The gram could, just as the ounce might, possibly satisfy all requirements in the U.S. system. For instance, we might say that an air traveler's luggage weighs 18,160 grams (40 pounds), but there are better ways of expressing weights or masses that are larger or smaller than 1 gram. These are explained in the next two sections.

How Is the Gram Multiplied for Heavier Weights (Larger Masses)?

We can express any heavier weight or larger mass as so many grams. But just as when ounces are used to express a very heavy weight, we can run into some cumbersome figures expressing many thousands of grams. Instead, we use decimal groupings of grams, naming each group as a new unit of measure containing that number of grams. Thus, 10 grams (g) = 1 decagram (dkg), 100 grams

(g) = 1 hectogram (hg), 1000 grams (g) = 1 kilogram (kg), 1,000,000 grams (g) = 1 metric ton (t or MT). See Table 1-1 in Chapter 1 for a list of the prefixes that are attached to the word *gram*.

The gram is a convenient unit for the measurement of average-size weights and masses, such as practical quantities of foods and other materials. Thus, 113½ grams of rice = ¼ pound. The kilogram is convenient for the measurement of larger weights and masses. Thus, 500 kilograms of water = 1102 pounds (about 132.3 gallons). A 100-kilogram quantity has a special name, the *quintal* (abbreviated *q*).

How Is the Gram Subdivided for Lighter Weights (Smaller Masses)?

Just as the gram is multiplied, as explained in the preceding section, it can also be subdivided into decimal groupings, each group being named as a new unit of measure corresponding to that decimal subdivision. Thus, 1/10 gram (g) = 1 decigram (dg), 1/100 gram (g) = 1 centigram (cg), 1/1000 gram (g) = 1 milligram (mg), 1/1,000,000 gram (g) = 1 microgram (μg).

The milligram is a convenient unit for the measurement of very small weights and masses (1 milligram = 0.0000353 ounce). A certain vitamin tablet, for example, is certified to contain 3.4 mg of iron.

The microgram is employed for extremely small weights and masses (1 microgram = 3.53×10^{-8} ounce). The same vitamin tablet cited in the preceding paragraph is certified to contain 10 μg of copper.

How Can Metric Units of Weight and Mass Be Converted to U.S. Units?

By means of fairly simple multiplications. In some cases, the multipliers contain several significant figures. The more of the figures you use, the more accurate the answer will be, although this makes longhand multiplication tedious. However, if you employ a small electronic calculator, long multiplication will offer no difficulty. Examples of common conversions follow (U.S. weights shown are avoirdupois). Others will be found in Appendix 2.

Table 5-1. Common Units of Weight

Unit	Abbreviation	Grams
Metric ton	Mt or t	1,000,000 grams
Quintal	q	100,000 grams
Kilogram	kg	1000 grams
Hectogram	hg	100 grams
Decagram	dkg	10 grams
Gram	g or gm	1 gram
Decigram	dg	1/10 gram
Centigram	cg	1/100 gram
Milligram	mg	1/1000 gram
Microgram	μg	1/1,000,000 gram

To convert	*Multiply by*
Grams to U.S. Ounces:	0.03527
Grams to U.S. Pounds:	0.0022
Kilograms to U.S. Ounces:	35.274
Kilograms to U.S. Pounds:	2.2046
Kilograms to U.S. Tons (Short):	0.0011
Metric Tons to U.S. Tons (Short):	1.1023
Metric Tons to U.S. Pounds:	2204.6
Milligrams to U.S. Ounces:	0.0000353
Milligrams to U.S. Pounds:	0.0000022
Milligrams to U.S. Grains:	0.01543
Micrograms to U.S. Ounces:	3.53×10^{-8}

Illustrative Example: How many U.S. ounces in 985 grams? Ounces = grams × 0.03527. So, 985 × 0.03527 = 34.74 ounces.

Illustrative Example: An American tourist weighing himself in Italy finds that the scale reads 71.2 kilograms. What is his weight in U.S. pounds? Pounds = kilograms × 2.2046.   So,   71.2 × 2.2046 = 156.97 pounds (approximately 157 lb).

Illustrative Example: A shipment to a processing mill consists of 50 metric tons of ore. How many U.S. tons (short) are in this shipment? U.S. short tons = metric tons × 1.1023. So, 50 × 1.1023 = 55.11 U.S. short tons.

Illustrative Example: The vitamin-C content of a certain multiple vitamin tablet is given on the label of the bottle as 100 milligrams. How many U.S. ounces of vitamin C does this represent? Ounces = milligrams × 0.0000353.   So,   100 × 0.0000353 = 0.00353 ounce.

Illustrative Example: How many U.S. ounces in 885 micrograms?   Ounces = micrograms × 3.35×10^{-8}. So, 885 × (3.53×10^{-8}) = 0.000031 ounce.

How Can U.S. Units of Weight and Mass Be Converted to Metric Units?

Again, by means of simple multiplications. As in the preceding case, however, some of the multipliers contain several figures which can make a longhand multiplication tedious. But, as mentioned before, a relatively inexpensive small electronic calculator can ease the task. Examples of common conversions follow (U.S. weights shown are avoirdupois). Others will be found in Appendix 2.

To convert	Multiply by
U.S. Ounces to Grams:	28.3495
U.S. Ounces to Kilograms:	0.02835
U.S. Ounces to Milligrams:	28,349
U.S. Ounces to Micrograms:	28,349,527
U.S. Grains to Milligrams:	64.798
U.S. Pounds to Grams:	453.59

U.S. Pounds to Kilograms:	0.45359
U.S. Pounds to Metric Tons:	0.0004536
U.S. Pounds to Milligrams:	453,592
U.S. Tons (Short) to Metric Tons:	0.90718
U.S. Tons (Short) to Kilograms:	907.18

Illustrative Example: How many grams in 8 U.S. ounces? Grams = ounces × 28.3495. So, 8 × 28.3495 = 226.796 grams.

Illustrative Example: A 5-gallon bottle of water has a net weight of 41.68 pounds. What is this weight in kilograms? Kilograms = pounds × 0.45359. So, 41.68 × 0.45359 = 18.90 kilograms.

Illustrative Example: A certain inventory contains 112 U.S. short tons of steel. How many metric tons? Metric tons = short tons × 0.90718. So, 112 × 0.90718 = 101.6 metric tons.

Illustrative Example: How many milligrams in a 5-grain aspirin tablet? Milligrams = grains × 64.798. So, 5 × 64.798 = 323.99 milligrams (approximately 324 mg).

PRACTICE EXERCISES

1. Convert 150 grams to U.S. ounces.
2. Convert 551 grams to U.S. pounds.
3. Convert 37 kilograms to U.S. ounces.
4. Convert ¾ kilogram to U.S. pounds.
5. Convert 1000 kilograms to U.S. short tons.
6. Convert 10 metric tons to U.S. short tons.
7. Convert 2 metric tons to U.S. pounds.
8. Convert 1225 milligrams to U.S. pounds.
9. Convert 900 milligrams to U.S. ounces.
10. Convert 110 milligrams to U.S. grains.
11. Convert 15 U.S. ounces to grams.
12. Convert 11½ U.S. ounces to kilograms.

13. Convert 2½ U.S. ounces to milligrams.
14. Convert 12 U.S. grains to milligrams.
15. Convert 22 U.S. pounds to grams.
16. Convert 83 U.S. pounds to kilograms.
17. Convert 2230 U.S. pounds to metric tons.
18. Convert ¼ U.S. pound to milligrams.
19. Convert ½ U.S. short ton to metric tons.
20. Convert 22¼ U.S. short tons to kilograms.

CHAPTER 6

Units of Temperature

The metric system originally set no standards for temperature measurement. However, the centigrade thermometer scale (invented in 1742 by the Swedish astronomer Anders Celsius) was immediately compatible with the metric system because it is a decimal-type scale. It seems highly probable—even necessary—that U.S. adoption of the metric system will also entail a switch from our cumbersome Fahrenheit thermometer scale to the centigrade scale, now called the Celsius scale in honor of its inventor.

The Fahrenheit scale is older than the Celsius, having been invented (for the mercury thermometer) in 1714 by Gabriel Daniel Fahrenheit, a German physicist. It has long been in everyday use in the United States, although the Celsius (centigrade) scale has been used almost exclusively for scientific purposes in this country.

This chapter is devoted to temperature units and their conversion.

How Do the Celsius and Fahrenheit Scales Compare?

On both scales (Fig. 6-1), the two key points are (1) the temperature at which water boils and (2) the temperature at which water freezes—both at an atmospheric pressure of 760 millimeters of mercury. On the Fahrenheit scale, the freezing point is designated 32 degrees ($32°F$), and the boiling point is 212 degrees ($212°F$). Thus, there are 180 equal degrees between the freezing point and the boiling point on the Fahrenheit scale. On the Celsius (centigrade) scale, the freezing point is designated zero degrees ($0°C$), and the boiling point is 100 degrees ($100°C$). Thus there are 100 equal degrees between the freezing point and the boiling point on the Celsius scale. It goes without saying that this latter arrangement, being decimal based, promotes ease of reading temperatures and of making calculations involving temperature.

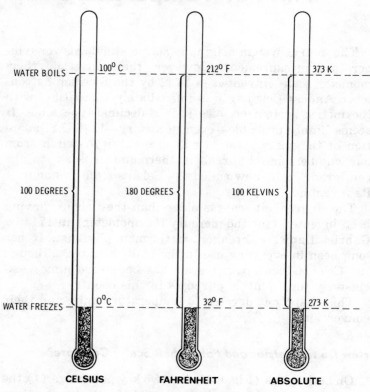

Fig. 6-1. Comparison of thermometer scales.

On the Fahrenheit scale, the zero-degree point represents the temperature of a mixture of equal parts (by weight) of snow and salt (sodium chloride). This temperature corresponds to 17.8 degrees below zero on the Celsius scale.

Table 6-1. Common Fahrenheit and Corresponding Celsius Temperatures

°F	°C	°F	°C
1000	537.8	40	4.4
500	260	32	0
212	100	20	−6.7
100	37.8	10	−12.2
90	32.2	0	−17.8
80	26.7	−5	−20.5
75	23.9	−10	−23.3
70	21.1	−15	−26.1
65	18.3	−20	−28.9
60	15.5	−30	−34.4
50	10		

How Is a Fahrenheit Reading Converted to Celsius?

By subtracting 32 from the Fahrenheit temperature and multiplying the result by the fraction 5/9. That is, °C = 5/9 (°F − 32).

Illustrative Example: Convert 75°F to Celsius.

$$°C = 5/9 (75 - 32) = 5/9 (43) = \frac{5 \times 43}{9} = \frac{215}{9} = 23.89°C.$$

Illustrative Example: A recipe calls for baking a fruit cake at an oven temperature of 275°F. What Celsius temperature does this correspond to?

$$°C = 5/9 (275 - 32) = 5/9 (243) = \frac{5 \times 243}{9} = \frac{1215}{9} = 135°C.$$

How Is a Celsius Reading Converted to Fahrenheit?

By multiplying the Celsius temperature by 9/5 and adding 32 to the result. That is, $°F = (9/5\ °C) + 32$.

Illustrative Example: Convert 85°C to Fahrenheit.
$°F = (9/5 \times 85) + 32 = \dfrac{9 \times 85}{5} + 32 = 765/5 + 32 =$
$153 + 32 = 185°F$.

Illustrative Example: The temperature of the glowing filament in an electric light bulb is given as 2800°C. What Fahrenheit temperature does this correspond to?
$°F = (9/5 \times 2800) + 32 = \dfrac{9 \times 2800}{5} + 32 =$
$\quad 25{,}200/5 + 32 = 5040 + 32 = 5072°F$.

What Is the Kelvin?

In the International System of Units (SI), the *kelvin* (abbreviated K) is the unit of temperature. This unit does not supplant the Celsius degree, but it will be more suitable in some types of work, especially scientific. Kelvins may be converted to degrees Celsius by subtracting 273.15. Thus, 1055 K = 1055 − 273.15 = 781.85°C. Conversely, degrees Celsius may be converted to kelvins by adding 273.15. Thus, 150°C = 150 + 273.15 = 423.15 K.

PRACTICE EXERCISES

1. Convert 98°F to °C.
2. Average body temperature is given as 98.6°F. Convert this temperature to °C.
3. How many kelvins in 3000°C?
4. Convert 450°C to °F.
5. Convert 21.1°C (ordinary room temperature) to °F.

Answers to Exercises

CHAPTER 2

1. 6.015 yards.
2. 46.6 miles.
3. 4.843 inches.
4. 4.33 inches.
5. 98.425 feet.
6. 1448.4 kilometers.
7. 5.029 meters.
8. 1.676 meters.
9. 30.48 centimeters.
10. 3.175 millimeters.

CHAPTER 3

1. 3861 square miles.
2. 10,763,870 square feet.
3. 96.874 square feet
4. 24.71 acres.
5. 7.749 square inches.
6. 284.9 square kilometers.
7. 12.14 hectares.
8. 27.80 square meters.
9. 13.378 square meters.
10. 322.58 square centimeters.

CHAPTER 4

1. 71.93 cubic yards.
2. 44,142.5 cubic feet.
3. 16.77 barrels.
4. 2.32 pints.
5. 0.871 gallon.
6. 5.07 fluid ounces.

7. 15.85 quarts.
8. 1.057 pint.
9. 34.33 gallons.
10. 2.54 fluid ounces.
11. 573.38 cubic meters.
12. 41.58 cubic meters.
13. 0.231 cubic meter.

14. 709.8 cubic centimeters.
15. 757.1 cubic centimeters.
16. 591.5 cubic centimeters.
17. 0.2366 liter.
18. 18.93 liters.
19. 14.79 milliliters.
20. 118.3 milliliters.

CHAPTER 5

1. 5.29 ounces.
2. 1.212 pounds.
3. 1305 ounces.
4. 1.653 pound.
5. 1.1 short ton.
6. 11.02 short tons.
7. 4409 pounds.
8. 0.00269 pound.
9. 0.0318 ounce.
10. 1.697 grain.

11. 425.2 grams.
12. 0.326 kilogram.
13. 70,872 milligrams.
14. 777.6 milligrams.
15. 9978.98 grams.
16. 37.65 kilograms.
17. 1.01 metric ton.
18. 113,398 milligrams.
19. 0.4536 metric ton.
20. 20,185 kilograms.

CHAPTER 6

1. 36.7°C.
2. 37°C.
3. 3273.15 K.

4. 842°F.
5. 69.98°F.

Conversion Factors

TO CONVERT FROM	TO	MULTIPLY BY
Acres	Ares	40.468726
Acres	Hectares	0.40468726
Acres	Sq cm	40,468,726
Acres	Sq km	0.00404687
Acres	Sq m	4046.8726
Acres	Sq mm	4.0468726×10^9
Acre-feet	Cu m	1233.49
Acre-feet/hr	Liters/sec	342.7
Acre-feet/min	Liters/sec	20,560
Ares	Acres	0.02471044
Ares	Hectares	0.01
Ares	Sq decameters	1
Ares	Sq m	100
Assay tons	Grams	29.1667
Atmospheres	Grams/sq cm	1033.228
Atmospheres	Kg/sq cm	1.033228
Atmospheres	Kg/sq m	10,332.28
Barrels (dry)	Cu m	0.11562
Barrels (liquid)	Cu m	0.11924
Bars	Grams/sq cm	1019.716

TO CONVERT FROM	TO	MULTIPLY BY
Bars	Kg/sq cm	1.019716
Bars	Kg/sq m	10,197.16
Baryes	Grams/sq cm	0.001019716
Bolts of cloth	Meters	36.576
Btu (IT)	Gram-cm	1.07588×10^7
Btu (IT)	Kg-m	107.588
Btu (IT)	Liter-atm	10.4125
Btu (mean)	Cu cm-atm	10,410.7
Btu (mean)	Gram-cm	1.07566×10^7
Btu (mean)	Kg-m	107.566
Btu (mean)	Liter-atm	10.4104
Btu (IT)/lb	Cu cm-atm/gram	22.9563
Btu (IT)/lb	Kg-m/gram	0.237191
Btu (mean)/lb	Kg-m/gram	0.237144
Btu (IT)/sec	Kg-m/sec	107.588
Btu (mean)/sec	Kg-m/sec	107.566
Bushels	Cu cm	35,239.3
Bushels	Cu m	0.0352393
Bushels	Decaliters	3.5238329
Bushels	Hectoliters	0.35238329
Bushels	Liters	35.238329
Cable lengths	Centimeters	21,945.6
Cable lengths	Kilometers	0.219456
Cable lengths	Meters	219.456
Calories, gram (IT)	Btu (IT)	0.00396832
Calories, gram (IT)	Calories, gram (mean)	1.00020
Calories, gram (IT)	Calories, kilogram (IT)	0.001
Calories, gram (IT)	Cu cm-atm	41.3214
Calories, gram (IT)	Gram-cm	42,694.3
Calories, gram (IT)	Kg-meters	0.426944

TO CONVERT FROM	TO	MULTIPLY BY
Calories, gram (IT)	Liter-atm	0.0413202
Calories, gram (mean)	Calories, gram (IT)	0.99980
Calories, gram (mean)	Calories, kg (mean)	0.001
Calories, gram (mean)	Cu cm-atm	41.3131
Calories, gram (mean)	Cu ft-atm	0.00145895
Calories, gram (mean)	Gram-cm	42,685.8
Calories, gram (mean)	Kg-meters	0.426858
Calories, gram (mean)	Liter-atm	0.0413119
Calories, kg (IT)	Btu (IT)	3.96832
Calories, kg (IT)	Btu (mean)	3.96911
Calories, kg (IT)	Calories, gram (IT)	1000
Calories, kg (IT)	Calories, kg (mean)	1.00020
Calories, kg (IT)	Cu cm-atm	41,321.4
Calories, kg (IT)	Cu ft-atm	1.45924
Calories, kg (IT)	Gram-cm	4.26943×10^7
Calories, kg (IT)	Kg-meters	426.943
Calories, kg (IT)	Liter-atm	41.3202
Calories, kg (mean)	Calories, gram (mean)	1000

TO CONVERT FROM	TO	MULTIPLY BY
Calories, kg (mean)	Calories, kg (IT)	0.99980
Calories, kg (mean)	Cu cm-atm	41,313.1
Calories, kg (mean)	Cu ft-atm	1.45895
Calories, kg (mean)	Gram-cm	4.26858×10^7
Calories, kg (mean)	Kg-meters	426.858
Calories, kg (mean)	Liter-atm	41.3119
Carats (metric)	Grains	3.08647
Carats (metric)	Grams	0.2
Carats (metric)	Milligrams	200
Centares	Ares	0.01
Centares	Sq feet	10.764
Centares	Sq inches	1550
Centares	Sq meters	1
Centares	Sq yards	1.196
Centigrams	Grains	0.15432356
Centigrams	Grams	0.01
Centiliters	Cubic cm	10.00027
Centiliters	Cubic in	0.61025
Centiliters	Drams (fluid)	2.705179
Centiliters	Liters	0.01
Centiliters	Ounces (fluid)	0.33815
Centimeters	Angstrom units	1×10^8
Centimeters	Feet	0.03280833
Centimeters	Inches	0.3937
Centimeters	Kilometers	1×10^{-5}
Centimeters	Meters	0.01
Centimeters	Microns	10,000
Centimeters	Miles (statute)	6.213699×10^{-6}
Centimeters	Millimeters	10
Centimeters	Mils	393.7
Centimeters	Rods	0.001988384

TO CONVERT FROM	TO	MULTIPLY BY
Centimeters	Yards	0.010936111
Centimeters/sec	Feet/min	1.9684998
Centimeters/sec	Feet/sec	0.03280833
Centimeters/sec	Kilometers/hr	0.036
Centimeters/sec	Kilometers/min	0.0006
Centimeters/sec	Meters/min	0.6
Centimeters/sec	Meters/sec	0.01
Centimeters/sec	Miles/hr	0.0223693
Centimeters/sec	Miles/min	0.000372822
Centimeters/sec	Miles/sec	6.21370×10^{-6}
Circular inches	Circ. mils	1,000,000
Circular inches	Sq cm	5.0671
Circular inches	Sq in.	0.78540
Circular inches	Sq mils	7.8540×10^{5}
Circular mm	Sq cm	0.0078540
Circular mm	Sq mm	0.78540
Circular mils	Sq cm	5.0671×10^{-6}
Circular mils	Sq in.	7.8540×10^{-7}
Circular mils	Sq mm	0.00050671
Circular mils	Sq mils	0.78540
Cords	Cu meters	3.625
Cubic cm	Bushels	2.83776×10^{-5}
Cubic cm	Cu decameters	1×10^{-9}
Cubic cm	Cu decimeters	0.001
Cubic cm	Cu ft	3.5314445×10^{-5}
Cubic cm	Cu in.	0.06102338
Cubic cm	Cu meters	1×10^{-6}
Cubic cm	Cu mm	1000
Cubic cm	Cu yards	1.3079428×10^{-6}
Cubic cm	Drams (fluid)	0.27053
Cubic cm	Gallon (dry)	0.00022702
Cubic cm	Gallons (liq.)	0.000264173
Cubic cm	Gills	0.00845351
Cubic cm	Liters	0.000999973
Cubic cm	Milliliters	0.999973
Cubic cm	Minims	16.231
Cubic cm	Ounces (fluid)	0.033814

TO CONVERT FROM	TO	MULTIPLY BY
Cubic cm	Pints (dry)	0.0018162
Cubic cm	Pints (liq.)	0.0021134
Cubic cm	Quarts (dry)	0.00090808
Cubic cm	Quarts (liq.)	0.0010567
Cubic cm/gram	Cubic ft/lb	0.0160185
Cubic cm/sec	Cubic ft/min	0.0021186
Cubic cm-atm	Btu (IT)	9.60356×10^{-5}
Cubic cm-atm	Btu (mean)	9.60548×10^{-5}
Cubic cm-atm	Cal, gram (IT)	0.0242006
Cubic cm-atm	Cal, gram (mean)	0.0242054
Cubic cm-atm	Cu ft-atm	3.5314445×10^{-5}
Cubic cm-atm	Foot-pounds	0.0747333
Cubic cm-atm	Hp-hr	3.77441×10^{-8}
Cubic cm-atm	Joules (Abs)	0.101325
Cubic cm-atm	Joules (Int)	0.101305
Cubic cm-atm	Kg-meters	0.0103323
Cubic cm-atm	Kw-hr (Abs)	2.81458×10^{-8}
Cubic cm-atm	Kw-hr (Int)	2.81402×10^{-8}
Cubic cm-atm	Watt-hr (Abs)	2.81458×10^{-5}
Cubic cm-atm	Watt-hr (Int)	2.81402×10^{-5}
Cubic decimeters	Cu cm	1000
Cubic decimeters	Cu decameters	0.000001
Cubic decimeters	Cu feet	0.035314445
Cubic decimeters	Cu inches	61.02338
Cubic decimeters	Cu meters	0.001
Cubic decimeters	Cu yards	0.00130794
Cubic decimeters	Liters	0.999973
Cubic feet	Cu cm	28,317.02
Cubic feet	Cu decimeters	28.31702

TO CONVERT FROM	TO	MULTIPLY BY
Cubic feet	Cu inches	1728
Cubic feet	Cu meters	0.02831702
Cubic feet	Cu yards	0.0370370
Cubic feet	Gallons (dry)	6.42851
Cubic feet	Gallons (liq.)	7.48052
Cubic feet	Kiloliters	0.02831625
Cubic feet	Liters	28.31625
Cubic feet	Ounces (fluid)	957.51
Cubic feet	Pints (liq.)	59.844
Cubic feet	Quarts (dry)	25.714
Cubic feet	Quarts (liq.)	29.922
Cubic ft/min	Cubic cm/sec	471.950
Cubic ft/min	Cubic ft/sec	0.016667
Cubic ft/min	Gallons/min	7.4805
Cubic ft/min	Gallons/sec	0.12468
Cubic ft/min	Liters/sec	0.47193
Cubic ft/sec	Gallons/hr	26,929.8
Cubic ft/sec	Gallons/min	448.831
Cubic ft/sec	Gallons/sec	7.4805
Cubic ft/sec	Liters/min	1698.96
Cubic ft/sec	Liters/sec	28.316
Cubic ft-atm	Btu (IT)	2.71944
Cubic ft-atm	Btu (mean)	2.71998
Cubic ft-atm	Cal, gram (IT)	685.288
Cubic ft-atm	Cal, gram (mean)	685.425
Cubic ft-atm	Kg-meters	292.580
Cubic hectometers	Cubic meters	1,000,000
Cubic inches	Centiliters	1.638673
Cubic inches	Cu cm	16.387162
Cubic inches	Cu decimeters	0.01638716
Cubic inches	Cubic ft	0.000578704
Cubic inches	Cu meters	1.638716×10^{-5}
Cubic inches	Cu mm	16,387.16
Cubic inches	Cu yard	2.143347×10^{-5}
Cubic inches	Decaliters	0.001638673
Cubic inches	Drams (fluid)	4.43290

TO CONVERT FROM	TO	MULTIPLY BY
Cubic inches	Gallons (dry)	0.00372020
Cubic inches	Gallons (liq.)	0.00432900
Cubic inches	Gills	0.138528
Cubic inches	Liters	0.01638673
Cubic inches	Milliliters	16.38673
Cubic inches	Minims	265.974
Cubic inches	Ounces (fluid)	0.55411
Cubic inches	Pints (dry)	0.0297616
Cubic inches	Pints (liq.)	0.0346320
Cubic inches	Quarts (dry)	0.0148808
Cubic inches	Quarts (liq.)	0.0173160
Cubic meters	Acre-feet	0.00081071
Cubic meters	Barrels (dry)	8.6490
Cubic meters	Barrels (liq.)	8.3865
Cubic meters	Bushels	28.3776
Cubic meters	Cu cm	1,000,000
Cubic meters	Cu decimeters	1000
Cubic meters	Cu ft	35.314445
Cubic meters	Cu hectometers	0.000001
Cubic meters	Cu inches	61,023.38
Cubic meters	Cu km	1×10^{-9}
Cubic meters	Cu mm	1×10^9
Cubic meters	Cu yards	1.3079428
Cubic meters	Decisteres	10
Cubic meters	Gallons (liq.)	264.173
Cubic meters	Hogsheads	4.1932
Cubic meters	Kiloliters	0.999973
Cubic meters	Liters	999.973
Cubic meters	Pints (liq.)	2113.4
Cubic meters	Quarts (liq.)	1056.7
Cubic meters	Steres	1
Cubic millimeters	Cu cm	0.001
Cubic millimeters	Cu inches	6.1023×10^{-5}
Cubic millimeters	Cu meters	1×10^{-9}

TO CONVERT FROM	TO	MULTIPLY BY
Cubic millimeters	Minims	0.016231
Cubic yards	Cu cm	764,559.45
Cubic yards	Cu decimeters	764.55945
Cubic yards	Cu feet	27
Cubic yards	Cu meters	0.76455945
Cubic yards	Kiloliters	0.76454
Cubic yards	Liters	764.54
Cubic yards	Ounces (fluid)	25,852.7
Decagrams	Grams	10
Decaliters	Bushels	0.28378
Decaliters	Cu inches	610.250
Decaliters	Liters	10
Decaliters	Pecks	1.13513
Decaliters	Pints (dry)	18.16204
Decaliters	Quarts (dry)	9.08102
Decameters	Meters	10
Decigrams	Grams	0.1
Decigrams	Kilograms	0.0001
Decigrams	Milligrams	100
Deciliters	Kiloliters	0.0001
Deciliters	Liters	0.1
Deciliters	Milliliters	100
Decimeters	Kilometers	0.0001
Decimeters	Meters	0.1
Decimeters	Millimeters	100
Decisteres	Cubic meters	0.1
Drams (apoth. or troy)	Grams	3.8879351
Drams (apoth. or troy)	Kilograms	0.0038879
Drams (apoth. or troy)	Milligrams	3887.9351
Drams (avdp)	Grams	1.771845
Drams (avdp)	Kilograms	0.001771845
Drams (avdp)	Milligrams	1771.845
Drams (fluid)	Cu cm	3.6967

TO CONVERT FROM	TO	MULTIPLY BY
Drams (fluid)	Liters	0.00369661
Drams (fluid)	Milliliters	3.69661
Drams (fluid)	Minims	60
Dyne-centimeters	Inch-pounds	8.85073×10^{-7}
Feet	Centimeters	30.48006096
Feet	Kilometers	0.0003048006
Feet	Meters	0.3048006096
Feet	Millimeters	304.8006096
Feet/hr	Cm/hr	30.48006
Feet/hr	Cm/min	0.508001
Feet/hr	Cm/sec	0.00846668
Feet/hr	Km/hr	0.0003048
Feet/hr	Km/min	5.08001×10^{-6}
Feet/hr	Km/sec	8.46668×10^{-8}
Feet/hr	Meters/hr	0.3048006
Feet/hr	Meters/min	0.00508001
Feet/hr	Meters/sec	8.46668×10^{-5}
Feet/min	Cm/sec	0.508001
Feet/min	Km/hr	0.0182880
Feet/min	Km/min	0.0003048
Feet/min	Meters/min	0.3048006
Feet/min	Meters/sec	0.00508001
Feet/second	Cm/sec	30.48006
Feet/second	Km/hr	1.09728
Feet/second	Km/min	0.0182880
Feet/second	Meters/min	18.2880
Feet/second	Meters/sec	0.3048006
Foot-candles	Lumens/sq meter	10.7639
Foot-lamberts	Candles/sq cm	0.000342624
Foot-poundals	Cal, gram (IT)	0.0100648
Foot-poundals	Cal, gram (mean)	0.0100669
Foot-poundals	Cu cm-atm	0.415892
Foot-poundals	Dyne-cm	4.21402×10^{5}
Foot-poundals	Gram-cm	429.711
Foot-poundals	Kg-meters	0.00429711
Foot-poundals	Liter-atm	0.000415880

TO CONVERT FROM	TO	MULTIPLY BY
Foot-pounds	Cal, gram (IT)	0.323826
Foot-pounds	Cal, gram (mean)	0.323891
Foot-pounds	Cal, Kg (IT)	0.000323826
Foot-pounds	Cal, kg (mean)	0.000323891
Foot-pounds	Cu cm-atm	13.3809
Foot-pounds	Dyne-cm	1.35582×10^7
Foot-pounds	Gram-cm	13,825.5
Foot-pounds	Kg-meters	0.138255
Foot-pounds	Liter-atm	0.0133805
Furlongs	Centimeters	20,117.2
Furlongs	Kilometers	0.201172
Furlongs	Meters	201.172
Gallons (dry)	Cubic cm	4404.9
Gallons (dry)	Liters	4.0476
Gallons (liq.)	Cubic cm	3785.434
Gallons (liq.)	Cu meters	0.00378543
Gallons (liq.)	Kiloliters	0.003785332
Gallons (liq.)	Liters	3.785332
Gallons (liq.)	Milliliters	3785.332
Gills	Cubic cm	118.2947
Gills	Liters	0.1182915
Gills	Milliliters	118.2915
Grains	Carats (metric)	0.32399
Grains	Grams	0.064798918
Grains	Kilograms	6.4798918×10^{-5}
Grains	Milligrams	64.798918
Grams	Decagrams	0.1
Grams	Decigrams	10
Grams	Drams (apoth or troy)	0.2572059
Grams	Drams (avdp)	0.5643833
Grams	Dynes	980.665
Grams	Grains	15.432356
Grams	Hectograms	0.01
Grams	Kilogram	0.001
Grams	Micrograms	1,000,000
Grams	Milligrams	1000

TO CONVERT FROM	TO	MULTIPLY BY
Grams	Myriagrams	0.0001
Grams	Ounces (apoth or troy)	0.03215074
Grams	Ounces (avdp)	0.03527396
Grams	Pennyweights	0.64301485
Grams	Poundals	0.0709315
Grams	Pounds (apoth or troy)	0.00267923
Grams	Pounds (avdp)	0.00220462
Grams	Scruples (apoth)	0.771618
Grams	Tons (long)	9.84206×10^{-7}
Grams	Tons (metric)	0.000001
Grams	Tons (short)	1.10231×10^{-6}
Grams/ centimeter	Dynes/cm	980.665
Grams/ centimeter	Poundals/in.	0.180166
Grams/ centimeter	Pounds/ft	0.067197
Grams/ centimeter	Pounds/in.	0.0055997
Grams/ centimeter	Pounds/mile	354.800
Grams/ centimeter	Pounds/yard	0.201591
Grams/ centimeter	Tons (long)/mile	0.15839
Grams/ centimeter	Tons (metric)/km	0.1
Grams/ centimeter	Tons (short)/mile	0.17740
Grams/cu cm	Dynes/cu cm	980.665
Grams/cu cm	Grains/cu ft	436,998
Grams/cu cm	Grains/milliliter	15.43277
Grams/cu cm	Grams/milliliter	1.000027
Grams/cu cm	Kg/cu meter	1000
Grams/cu cm	Kg/hectoliter	100.0027

TO CONVERT FROM	TO	MULTIPLY BY
Grams/cu cm	Kg/liter	1.000027
Grams/cu cm	Poundals/cu in.	1.162365
Grams/cu cm	Pounds/cu ft	62.4282595
Grams/cu cm	Pounds/cu in.	0.036127465
Grams/cu cm	Pounds/cu yard	1685.5630
Grams/cu cm	Pounds/gal (dry)	9.71106
Grams/cu cm	Pounds/gal (liq.)	8.34545
Grams/cu cm	Tons (long)/ cu yd	0.75248
Grams/cu cm	Tons (metric)/ cu meter	1
Grams/cu cm	Tons (short)/ cu yd	0.8427815
Grams/liter	Grains/gal	58.417
Grams/liter	Grams/cu cm	0.000999973
Grams/liter	Pounds/cu ft	0.062427
Grams/milliliter	Grams/cu cm	0.999973
Grams/sq cm	Atmospheres	0.000967841
Grams/sq cm	Bars	0.000980665
Grams/sq cm	Dynes/sq cm	980.665
Grams/sq cm	Kg/sq meter	10
Grams/sq cm	Poundals/sq in.	0.457620
Grams/sq cm	Pounds/sq ft	2.04816
Grams/sq cm	Pounds/sq in.	0.01422333
Gram-cm	Btu (IT)	9.29472×10^{-8}
Gram-cm	Btu (mean)	9.29658×10^{-8}
Gram-cm	Cal, gram (IT)	2.34223×10^{-5}
Gram-cm	Cal, gram (mean)	2.34270×10^{-5}
Gram-cm	Cal, Kg (IT)	2.34223×10^{-8}
Gram-cm	Cal, Kg (mean)	2.34270×10^{-8}
Gram-cm	Ergs	980.665
Gram-cm	Foot-poundals	0.00232714
Gram-cm	Foot-pounds	7.23300×10^{-5}
Gram-cm	Kg-meters	0.00001
Hectares	Acres	2.471044
Hectares	Ares	100
Hectares	Centares	10,000

TO CONVERT FROM	TO	MULTIPLY BY
Hectares	Sq cm	1×10^8
Hectares	Sq ft	107,638.7
Hectares	Sq in.	15,499,969
Hectares	Sq km	0.01
Hectares	Sq meters	10,000
Hectares	Sq miles	0.003861006
Hectares	Sq rods	395.367
Hectares	Sq yards	11,959.85
Hectograms	Drams (avdp)	56.43833
Hectograms	Dynes	98,066.5
Hectograms	Grains	1543.2356
Hectograms	Grams	100
Hectograms	Kilograms	0.1
Hectograms	Ounces (apoth or troy)	3.215074
Hectograms	Ounces (avdp)	3.527396
Hectograms	Poundals	7.09315
Hectograms	Pounds (apoth or troy)	0.267923
Hectograms	Pounds (avdp)	0.220462
Hectoliters	Bushels	2.837819
Hectoliters	Cu cm	1.000027×10^5
Hectoliters	Cu decimeters	100.0027
Hectoliters	Cu ft	3.531539
Hectoliters	Cu in.	6102.50
Hectoliters	Cu meters	0.1000027
Hectoliters	Cu yards	0.130798
Hectoliters	Decaliters	10
Hectoliters	Drams (fluid)	27,051.8
Hectoliters	Gallons (dry)	22.7026
Hectoliters	Gallons (liq.)	26.4178
Hectoliters	Gills	845.368
Hectoliters	Kiloliters	0.1
Hectoliters	Microliters	1×10^8
Hectoliters	Milliliters	100,000
Hectoliters	Ounces (fluid)	3381.47
Hectoliters	Pecks	11.3513

TO CONVERT FROM	TO	MULTIPLY BY
Hectoliters	Pints (dry)	181.620
Hectoliters	Pints (liq.)	211.342
Hectoliters	Quarts (dry)	90.8102
Hectoliters	Quarts (liq.)	105.671
Hectometers	Centimeters	10,000
Hectometers	Feet	328.0833
Hectometers	Furlongs	0.4970959
Hectometers	Inches	3937
Hectometers	Kilometers	0.1
Hectometers	Megameters	0.0001
Hectometers	Meters	100
Hectometers	Microns	1×10^8
Hectometers	Miles	0.06213699
Hectometers	Millimeters	100,000
Hectometers	Millimicrons	1×10^{11}
Hectometers	Myriameters	0.01
Hectometers	Rods	19.88384
Hectometers	Yards	109.3611
Hogsheads	Cu meters	0.2384824
Hogsheads	Liters	238.4759
Hundredweights (long)	Kilograms	50.80235
Hundredweights (long)	Quintals	0.5080235
Hundredweights (short)	Kilograms	45.359243
Hundredweights (short)	Metric tons	0.04535924
Inches	Centimeters	2.54000508
Inches	Kilometers	$2.54000508 \times 10^{-5}$
Inches	Meters	0.0254000508
Inches	Microns	25,400.0508
Inches	Millimeters	25.4000508
Inches	Millimicrons	2.54000508×10^7
Kilograms	Decagrams	100

TO CONVERT FROM	TO	MULTIPLY BY
Kilograms	Drams (apoth or troy)	257.2059
Kilograms	Drams (avdp)	564.3833
Kilograms	Dynes	980,665
Kilograms	Grains	15,432.356
Kilograms	Grams	1000
Kilograms	Hundredweights (long)	0.019684128
Kilograms	Hundredweights (short)	0.02204622341
Kilograms	Milligrams	1,000,000
Kilograms	Myriagrams	0.1
Kilograms	Ounces (apoth or troy)	32.150742
Kilograms	Ounces (avdp)	35.27396
Kilograms	Pennyweights	643.01485
Kilograms	Poundals	70.93152
Kilograms	Pounds (apoth or troy)	2.6792285
Kilograms	Pounds (avdp)	2.204622341
Kilograms	Quintals	0.01
Kilograms	Scruples (apoth)	771.6178
Kilograms	Tons (long)	0.0009842064
Kilograms	Tons (metric)	0.001
Kilograms	Tons (short)	0.0011023112
Kilograms/cu meter	Grams/cu cm	0.001
Kilograms/cu meter	Kg/liter	0.001000027
Kilograms/cu meter	Lb/cu ft	0.0624283
Kilograms/cu meter	Lb/cu in.	3.61275×10^{-5}
Kilograms/cu meter	Lb/cu yd	1.68556
Kilograms/cu meter	Lb/gal (liq.)	0.00834545

TO CONVERT FROM	TO	MULTIPLY BY
Kilograms/ cu meter	Tons (long)/ cu yd	0.00075248
Kilograms/ cu meter	Tons (metric)/ cu meter	0.001
Kilograms/ cu meter	Tons (short)/ cu yd	0.00084278
Kilograms/sq cm	Atmospheres	0.967841
Kilograms/sq cm	Bars	0.980665
Kilograms/sq cm	Dynes/sq cm	980,665
Kilograms/sq cm	Grams/sq cm	1000
Kilograms/sq cm	Pounds/sq ft	2048.155
Kilograms/sq cm	Pounds/sq in.	14.2233
Kilograms/sq cm	Tons (short)/ sq ft	1.024078
Kilograms/sq cm	Tons (short)/ sq in.	0.00711165
Kilograms/ sq meter	Atmospheres	9.67841×10^{-5}
Kilograms/ sq meter	Bars	9.80665×10^{-5}
Kilograms/ sq meter	Dynes/sq cm	98.0665
Kilograms/ sq meter	Grams/sq cm	0.1
Kilograms/ sq meter	Pounds/sq ft	0.2048155
Kilograms/ sq meter	Pounds/sq in.	0.00142233
Kilograms/ sq meter	Tons (short)/ sq ft	0.0001024078
Kilograms/ sq meter	Tons (short)/ sq in.	7.11165×10^{-7}
Kilograms/sq mm	Atmospheres	96.7841
Kilograms/sq mm	Bars	98.0665
Kilograms/sq mm	Dynes/sq cm	9.80665×10^{7}
Kilograms/sq mm	Grams/sq cm	100,000
Kilograms/sq mm	Pounds/sq ft	2.048155×10^{5}

TO CONVERT FROM	TO	MULTIPLY BY
Kilograms/sq mm	Pounds/sq in.	1422.33
Kilograms/sq mm	Tons (short)/ sq ft	102.4078
Kilograms/sq mm	Tons (short)/ sq in.	0.711165
Kilogram-meters	Btu (IT)	0.00929472
Kilogram-meters	Btu (mean)	0.00929658
Kilogram-meters	Cal, gram (IT)	2.34223
Kilogram-meters	Cal, gram (mean)	2.34270
Kilogram-meters	Cal, kg (IT)	0.00234223
Kilogram-meters	Cal, kg (mean)	0.00234270
Kilogram-meters	Cu cm-atm	96.7842
Kilogram-meters	Cu ft-atm	0.00341788
Kilogram-meters	Dyne-cm	9.80665×10^7
Kilogram-meters	Ergs	9.80665×10^7
Kilogram-meters	Foot-poundals	232.715
Kilogram-meters	Foot-pounds	7.23300
Kilogram-meters	Gram-cm	100,000
Kilogram-meters	Hp-hr (U.S.)	3.65303×10^{-6}
Kilogram-meters	Hp-hr (metric)	3.70370×10^{-6}
Kilogram-meters	Liter-atm	0.0967816
Kiloliters	Bushels	28.37819
Kiloliters	Cu cm	1.000027×10^6
Kiloliters	Cu feet	35.31539
Kiloliters	Cu inches	61,025.0
Kiloliters	Cu meters	1.000027
Kiloliters	Cu yards	1.30798
Kiloliters	Decaliters	100
Kiloliters	Gallons (dry)	227.03
Kiloliters	Gallons (liq.)	264.178
Kiloliters	Hectoliters	10
Kiloliters	Liters	1000
Kiloliters	Milliliters	1,000,000
Kiloliters	Ounces (fluid)	33,814.7
Kiloliters	Pecks	113.513
Kiloliters	Pints (dry)	1816.204
Kiloliters	Pints (liq.)	2113.42

TO CONVERT FROM	TO	MULTIPLY BY
Kiloliters	Quarts (dry)	908.102
Kiloliters	Quarts (liq.)	1056.71
Kilometers	Centimeters	100,000
Kilometers	Feet	3280.833
Kilometers	Furlong	4.970959
Kilometers	Hectometers	10
Kilometers	Leagues	0.207123
Kilometers	Megameters	0.001
Kilometers	Meters	1000
Kilometers	Miles	0.6213699
Kilometers	Myriameters	0.1
Kilometers	Rods	198.8384
Kilometers	Yards	1093.611
Kilometers/hr	Cm/sec	27.77778
Kilometers/hr	Feet/hr	3280.833
Kilometers/hr	Feet/min	54.6806
Kilometers/hr	Feet/sec	0.911343
Kilometers/hr	Km/min	0.0166667
Kilometers/hr	Km/sec	0.000277778
Kilometers/hr	Meters/hr	1000
Kilometers/hr	Meters/min	16.6667
Kilometers/hr	Meters/sec	0.277778
Kilometers/hr	Miles/hr	0.6213699
Kilometers/hr	Miles/min	0.0103562
Kilometers/hr	Miles/sec	0.000172603
Kilometers/min	Cm/sec	1666.67
Kilometers/min	Feet/min	3280.84
Kilometers/min	Feet/sec	54.6806
Kilometers/min	Km/hr	60
Kilometers/min	Meters/min	1000
Kilometers/min	Meters/sec	16.6667
Kilometers/min	Miles/hr	37.28219
Kilometers/min	Miles/min	0.6213699
Leagues	Kilometers	4.82804
Light years	Kilometers	9.46091×10^{12}
Liters	Bushels	0.02837819
Liters	Cu cm	1000.027

TO CONVERT FROM	TO	MULTIPLY BY
Liters	Cu decimeters	1.000027
Liters	Cu feet	0.03531539
Liters	Cu inches	61.0250
Liters	Cu meters	0.001000027
Liters	Cu yards	0.00130798
Liters	Decaliters	0.1
Liters	Drams (fluid)	270.518
Liters	Gallons (dry)	0.227026
Liters	Gallons (liq.)	0.264178
Liters	Gills	8.45368
Liters	Hectoliters	0.01
Liters	Kiloliters	0.001
Liters	Microliters	1,000,000
Liters	Milliliters	1000
Liters	Minims	16,231.1
Liters	Ounces (fluid)	33.8147
Liters	Pecks	0.113513
Liters	Pints (dry)	1.81620
Liters	Pints (liq.)	2.11342
Liters	Quarts (dry)	0.908102
Liters	Quarts (liq.)	1.05671
Liters-atm	Btu (IT)	0.0960382
Liters-atm	Btu (mean)	0.0960574
Liters-atm	Cal, gram (IT)	24.2013
Liters-atm	Cal, gram (mean)	24.2061
Liters-atm	Cal, kg (IT)	0.0242013
Liters-atm	Cal, kg (mean)	0.0242061
Liters-atm	Cu cm-atm	1000.027
Liters-atm	Cu ft-atm	0.0353153
Liters-atm	Foot-poundals	2404.53
Liters-atm	Foot-pounds	74.7353
Liters-atm	Hp-hr	3.77451×10^{-5}
Liters-atm	Kg-meters	10.3326
Lumens/sq meter	Foot-candles	0.092903
Lumens/sq meter	Lumens/sq ft	0.092903
Lumens/sq meter	Lux	1
Lumens/sq meter	Meter-candles	1

TO CONVERT FROM	TO	MULTIPLY BY
Lumens/sq meter	Phots	0.0001
Lux	Lumens/sq meter	1
Lux	Meter-candles	1
Megameters	Meters	1,000,000
Meters	Bolts of cloth	0.027340
Meters	Centimeters	100
Meters	Fathoms	0.546806
Meters	Feet	3.280833
Meters	Furlongs	0.004970959
Meters	Hectometers	0.01
Meters	Inches	39.37
Meters	Kilometers	0.001
Meters	Megameters	0.000001
Meters	Micromicrons	1×10^{12}
Meters	Microns	1,000,000
Meters	Miles	0.0006213699
Meters	Millimeters	1000
Meters	Millimicrons	1×10^{9}
Meters	Mils	39,370
Meters	Myriameters	0.0001
Meters	Rods	0.1988384
Meters	Yards	1.093611
Meters/hr	Cm/hr	100
Meters/hr	Cm/min	1.66667
Meters/hr	Cm/sec	0.0277778
Meters/hr	Feet/hr	3.280833
Meters/hr	Feet/min	0.0546806
Meters/hr	Feet/sec	0.000911343
Meters/hr	Kilometers/hr	0.001
Meters/hr	Kilometers/min	1.66667×10^{-5}
Meters/hr	Kilometers/sec	2.77778×10^{-7}
Meters/hr	Meters/min	0.0166667
Meters/hr	Meters/sec	0.000277778
Meters/hr	Miles/hr	0.0006213699
Meters/hr	Miles/min	1.03562×10^{-5}
Meters/hr	Miles/sec	1.72603×10^{-7}
Meters/min	Cm/sec	1.66667

TO CONVERT FROM	TO	MULTIPLY BY
Meters/min	Feet/hr	196.850
Meters/min	Feet/min	3.280833
Meters/min	Feet/sec	0.0546806
Meters/min	Kilometers/hr	0.06
Meters/min	Kilometers/min	0.001
Meters/min	Meters/sec	0.0166667
Meters/min	Miles/hr	0.0372822
Meters/min	Miles/min	0.000621369
Meters/sec	Cm/sec	100
Meters/sec	Feet/min	196.850
Meters/sec	Feet/sec	3.280833
Meters/sec	Kilometers/hr	3.6
Meters/sec	Kilometers/min	0.06
Meters/sec	Meters/min	60
Meters/sec	Miles/hr	2.236932
Meters/sec	Miles/min	0.0372822
Micrograms	Grams	0.000001
Micrograms	Milligrams	0.001
Microhm-inches	Microhm-cm	2.540005
Microliters	Liters	0.000001
Micromicrons	Centimeters	1×10^{-10}
Micromicrons	Meters	1×10^{-12}
Micromicrons	Microns	0.000001
Microns	Centimeters	0.0001
Microns	Feet	3.280833×10^{-6}
Microns	Inches	3.937×10^{-5}
Microns	Kilometers	1×10^{-9}
Microns	Megameters	1×10^{-12}
Microns	Meters	0.000001
Microns	Millimeters	0.001
Microns	Millimicrons	1000
Microns	Mils	0.03937
Microns	Yards	1.093611×10^{-6}
Miles	Centimeters	160,934.72
Miles	Kilometers	1.6093472
Miles	Meters	1609.3472
Miles	Millimeters	1,609,347.2

TO CONVERT FROM	TO	MULTIPLY BY
Miles	Myriameters	0.16093472
Miles/hr	Cm/sec	44.704088
Miles/hr	Km/hr	1.6093472
Miles/hr	Km/min	0.02682245
Miles/hr	Meters/min	26.82245
Miles/hr	Meters/sec	0.4470409
Miles/min	Cm/sec	2682.245
Miles/min	Km/hr	96.56083
Miles/min	Km/min	1.609347
Miles/min	Meters/min	1609.347
Miles/min	Meters/sec	26.82245
Millibars	Dynes/sq cm	1000
Millibars	Grams/sq cm	1.019716
Milligrams	Carats (metric)	0.005
Milligrams	Drams (apoth or troy)	0.0002572059
Milligrams	Drams (avdp)	0.0005643833
Milligrams	Grains	0.015432356
Milligrams	Grams	0.001
Milligrams	Kilograms	0.000001
Milligrams	Micrograms	1000
Milligrams	Ounces (apoth or troy)	3.215074×10^{-5}
Milligrams	Ounces (avdp)	3.527396×10^{-5}
Milligrams	Pennyweights	0.000643015
Milligrams	Pounds (apoth or troy)	2.67923×10^{-6}
Milligrams	Pounds (avdp)	2.20462×10^{-6}
Milligrams	Scruples (apoth)	0.000771618
Milligrams/cm	Dynes/cm	0.980665
Milligrams/cm	Dynes/in.	2.49089
Milligrams/cm	Grains/in.	0.0391983
Milligrams/cm	Grams/cm	0.001
Milligrams/cm	Grams/in.	0.002540005
Milligrams/cm	Kg/kilometer	0.1
Milligrams/cm	Pounds/ft	0.000067197
Milligrams/cm	Pounds/in.	5.5997×10^{-6}

TO CONVERT FROM	TO	MULTIPLY BY
Milligrams/in.	Dynes/cm	0.386088
Milligrams/in.	Dynes/in.	0.980665
Milligrams/in.	Grains/in.	0.015432356
Milligrams/in.	Grams/cm	0.0003937
Milligrams/in.	Grams/in.	0.001
Milligrams/in.	Kg/kilometer	0.03937
Milligrams/in.	Pounds/ft	0.0000264554
Milligrams/in.	Pounds/in.	2.20462×10^{-6}
Milligrams/liter	Grains/gal	0.058417
Milligrams/liter	Grams/cu cm	9.99973×10^{-7}
Milligrams/liter	Grams/liter	0.001
Milligrams/liter	Pounds/cu ft	0.000062427
Milligrams/mm	Dynes/cm	9.80665
Milligrams/mm	Dynes/in.	24.9089
Milligrams/mm	Grains/in.	0.391983
Milligrams/mm	Grams/cm	0.01
Milligrams/mm	Grams/in.	0.02540005
Milligrams/mm	Kg/kilometer	1
Milligrams/mm	Pounds/ft	0.00067197
Milligrams/mm	Pounds/in.	0.000055997
Milliliters	Cu cm	1.000027
Milliliters	Cu inches	0.0610250
Milliliters	Drams (fluid)	0.270518
Milliliters	Gallons	0.000264178
Milliliters	Gills	0.00845368
Milliliters	Liters	0.001
Milliliters	Microliters	1000
Milliliters	Minims	16.2311
Milliliters	Ounces (fluid)	0.0338147
Milliliters	Pints (liq.)	0.00211342
Milliliters	Quarts (liq.)	0.00105671
Millimeters	Centimeters	0.1
Millimeters	Feet	0.003280833
Millimeters	Inches	0.03937
Millimeters	Kilometers	0.000001
Millimeters	Megameters	1×10^{-9}
Millimeters	Meters	0.001

TO CONVERT FROM	TO	MULTIPLY BY
Millimeters	Micromicrons	1×10^9
Millimeters	Microns	1000
Millimeters	Miles	6.213699×10^{-7}
Millimeters	Millimicrons	1,000,000
Millimeters	Mils	39.37
Millimeters	Rods	0.0001988384
Millimeters	Yards	0.001093611
Millimicrons	Centimeters	1×10^{-7}
Millimicrons	Feet	3.280833×10^{-9}
Millimicrons	Inches	3.937×10^{-8}
Millimicrons	Kilometers	1×10^{-12}
Millimicrons	Meters	1×10^{-9}
Millimicrons	Microns	0.001
Millimicrons	Millimeters	0.000001
Millimicrons	Mils	3.937×10^{-5}
Millimicrons	Yards	1.093611×10^{-9}
Milliphots	Meter-candles	10
Milliphots	Phots	0.001
Mils	Centimeters	0.00254000508
Mils	Feet	8.33333×10^{-5}
Mils	Inches	0.001
Mils	Kilometers	$2.54000508 \times 10^{-8}$
Mils	Microns	25.4000508
Mils	Millimeters	0.0254000508
Mils	Yards	2.77778×10^{-5}
Minims (fluid)	Cu cm	0.0616119
Minims (fluid)	Cu in.	0.00375977
Minims (fluid)	Cu mm	61.6119
Minims (fluid)	Drams (fluid)	0.0166667
Minims (fluid)	Gallons (liq.)	1.62760×10^{-5}
Minims (fluid)	Gills	0.000520833
Minims (fluid)	Liters	6.16102×10^{-5}
Minims (fluid)	Milliliters	0.0616102
Minims (fluid)	Ounces (fluid)	0.00208333
Minims (fluid)	Pints (liq.)	0.000130208
Minims (fluid)	Quarts (liq.)	6.5104×10^{-5}
Myriagrams	Grams	10,000

TO CONVERT FROM	TO	MULTIPLY BY
Myriagrams	Kilograms	10
Myriagrams	Pounds (avdp)	22.0462
Myriameters	Kilometers	10
Myriameters	Meters	10,000
Myriameters	Miles	6.21370
Ohm-cm	Ohm-in.	0.3937
Ohm-in.	Ohm-cm	2.540005
Ounces (apoth or troy)	Drams (apoth or troy)	8
Ounces (apoth or troy)	Grams	31.103481
Ounces (apoth or troy)	Kilograms	0.03110348
Ounces (apoth or troy)	Milligrams	31,103.481
Ounces (apoth or troy)	Tons (metric)	3.110349×10^{-5}
Ounces (avdp)	Grams	28.349527
Ounces (avdp)	Hectograms	0.28349527
Ounces (avdp)	Kilograms	0.028349527
Ounces (avdp)	Milligrams	28,349.527
Ounces (fluid)	Cu cm	29.5737
Ounces (fluid)	Cu ft	0.00104438
Ounces (fluid)	Cu in.	1.80469
Ounces (fluid)	Cu meters	2.95737×10^{-5}
Ounces (fluid)	Cu yards	3.86808×10^{-5}
Ounces (fluid)	Drams (fluid)	8
Ounces (fluid)	Gallons (dry)	0.00671380
Ounces (fluid)	Gallons (liq.)	0.0078125
Ounces (fluid)	Gills	0.25
Ounces (fluid)	Liters	0.0295729
Ounces (fluid)	Milliliters	29.5729
Ounces (fluid)	Minims	480
Ounces (fluid)	Pints (liq.)	0.0625
Ounces (fluid)	Quarts (liq.)	0.03125
Ounces/sq ft	Dynes/sq cm	29.92508
Ounces/sq ft	Grams/sq cm	0.0305151

TO CONVERT FROM	TO	MULTIPLY BY
Ounces/sq in.	Dynes/sq cm	4309.21
Ounces/sq in.	Grams/sq cm	4.39417
Pecks	Cu cm	8809.818
Pecks	Decaliters	0.880958
Pecks	Liters	8.80958
Pennyweights	Grams	1.5551740
Pennyweights	Kilograms	0.0015551740
Pennyweights	Milligrams	1555.1740
Phots	Lumens/sq cm	1
Phots	Lumens/ sq meter	10,000
Pints (dry)	Cu cm	550.6136
Pints (dry)	Decaliters	0.0550599
Pints (dry)	Liters	0.5505991
Pints (liq.)	Cu cm	473.1798
Pints (liq.)	Cu meters	0.0004731798
Pints (liq.)	Drams (fluid)	128
Pints (liq.)	Gallons (liq.)	0.125
Pints (liq.)	Liters	0.473167
Pints (liq.)	Milliliters	473.167
Poundals	Dynes	13,825.5
Poundals	Grams	14.0981
Poundals	Kilograms	0.0140981
Pounds (apoth or troy)	Grams	373.24177
Pounds (apoth or troy)	Kilograms	0.37324177
Pounds (apoth or troy)	Milligrams	373,241.77
Pounds (apoth or troy)	Tons (metric)	0.000373242
Pounds (avdp)	Grams	453.592423
Pounds (avdp)	Kilograms	0.453592423
Pounds (avdp)	Milligrams	453,592.423
Pounds (avdp)	Tons (metric)	0.0004535924
Pounds/cu ft	Grams/cu cm	0.01601837
Pounds/cu ft	Grams/liter	16.01880

TO CONVERT FROM	TO	MULTIPLY BY
Pounds/cu ft	Kg/cu meter	16.01837
Pounds/cu ft	Kg/hectoliter	1.601880
Pounds/cu ft	Tons (metric)/ cu meter	0.01601837
Pounds/cu in.	Grams/cu cm	27.67974
Pounds/cu in.	Grams/liter	27,680.49
Pounds/cu in.	Kg/cu meter	27,679.74
Pounds/cu in.	Kg/hectoliter	2768.049
Pounds/cu in.	Tons (metric)/ cu meter	27.6797
Pounds/cu yd	Grams/cu cm	0.00059327
Pounds/cu yd	Kg/cu meter	0.5932730
Pounds/cu yd	Kg/hectoliter	0.05932890
Pounds/foot	Grams/cm	14.88161
Pounds/foot	Grams/ft	453.5924
Pounds/foot	Grams/in.	37.79937
Pounds/foot	Kg/foot	0.4535924
Pounds/foot	Kg/kilometer	1488.161
Pounds/foot	Kg/meter	1.488161
Pounds/foot	Ounces/cm	0.5249333
Pounds/foot	Pounds/meter	3.280833
Pounds/foot	Tons (metric)/ft	0.0004535924
Pounds/foot	Tons (metric)/in.	3.779937×10^{-5}
Pounds/foot	Tons (metric)/km	1.488161
Pounds/foot	Tons (metric)/ meter	0.001488161
Pounds/foot	Tons (metric)/ mile	2.394968
Pounds/foot	Tons (metric)/yd	0.001360777
Pounds/foot	Tons (short)/km	1.640417
Pounds/inch	Grains/cm	2755.9
Pounds/inch	Grams/cm	178.5793
Pounds/inch	Grams/ft	5443.109
Pounds/inch	Grams/in.	453.5924
Pounds/inch	Kg/foot	5.443109
Pounds/inch	Kg/kilometer	17,857.93
Pounds/inch	Kg/meter	17.85793

TO CONVERT FROM	TO	MULTIPLY BY
Pounds/inch	Ounces/cm	6.2992
Pounds/inch	Pounds/meter	39.37
Pounds/inch	Tons (long)/km	17.57589
Pounds/inch	Tons (metric)/ft	0.005443109
Pounds/inch	Tons (metric)/in.	0.00045359
Pounds/inch	Tons (metric)/km	17.85793
Pounds/inch	Tons (metric)/ meter	0.01785793
Pounds/inch	Tons (metric)/ mile	28.73962
Pounds/inch	Tons (metric)/yd	0.01632933
Pounds/inch	Tons (short)/km	19.685
Pounds/sq foot	Dynes/sq cm	478.803
Pounds/sq foot	Grams/sq cm	0.488241
Pounds/sq foot	Kg/sq cm	0.000488241
Pounds/sq foot	Kg/sq meter	4.88241
Pounds/sq inch	Dynes/sq cm	68,947.6
Pounds/sq inch	Grams/sq cm	70.3067
Pounds/sq inch	Kg/sq cm	0.0703067
Pounds/sq inch	Kg/sq meter	703.067
Quarts (dry)	Cu cm	1101.227
Quarts (dry)	Decaliters	0.1101198
Quarts (dry)	Liters	1.101198
Quarts (liq.)	Cu cm	946.3596
Quarts (liq.)	Cu meters	0.000946359
Quarts (liq.)	Liters	0.946333
Quarts (liq.)	Milliliters	946.333
Quintals (metric)	Grams	100,000
Quintals (metric)	Kilograms	100
Quintals (metric)	Pounds (avdp)	220.4622
Rods	Centimeters	502.9210
Rods	Hectometers	0.0502921
Rods	Kilometers	0.00502921
Rods	Meters	5.029210
Scruples (apoth)	Grams	1.2959784
Scruples (apoth)	Kilograms	0.0012959784

TO CONVERT FROM	TO	MULTIPLY BY
Scruples (apoth)	Milligrams	1295.9784
Square cm	Acres	2.471044×10^{-8}
Square cm	Ares	0.000001
Square cm	Circ in.	0.197352
Square cm	Circ mm	127.324
Square cm	Circ mils	197,352
Square cm	Hectares	1×10^{-8}
Square cm	Sq decimeters	0.01
Square cm	Sq feet	0.001076387
Square cm	Sq inches	0.15499969
Square cm	Sq kilometers	1×10^{-10}
Square cm	Sq meters	0.0001
Square cm	Sq miles	3.861006×10^{-11}
Square cm	Sq mm	100
Square cm	Sq mils	154,999.7
Square cm	Sq rods	3.95367×10^{-6}
Square cm	Sq yards	0.00011959
Square cm/dyne	Sq cm/gram	980.665
Square cm/dyne	Sq cm/kg	980,665.0
Square cm/dyne	Sq ft/lb	478.80
Square cm/dyne	Sq in./lb	68,947
Square cm/gram	Sq cm/dyne	0.001019716
Square cm/kg	Sq cm/dyne	1.019716×10^{-6}
Sq decameters	Acres	0.02471044
Sq decameters	Ares	1
Sq decameters	Hectares	0.01
Sq decameters	Sq cm	1,000,000
Sq decameters	Sq feet	1076.387
Sq decameters	Sq inches	154,999.69
Sq decameters	Sq meters	100
Sq decameters	Sq yards	119.5992
Sq decimeters	Sq cm	100
Sq decimeters	Sq feet	0.1076387
Sq decimeters	Sq inches	15.499969
Sq decimeters	Sq kilometers	1×10^{-8}
Sq decimeters	Sq meters	0.01
Sq decimeters	Sq mm	10,000

TO CONVERT FROM	TO	MULTIPLY BY
Square feet	Ares	0.0009290341
Square feet	Hectares	9.290341×10^{-6}
Square feet	Sq cm	929.0341
Square feet	Sq kilometers	9.290341×10^{-8}
Square feet	Sq meters	0.09290341
Square feet	Sq mm	92,903.41
Square hectometers	Acres	2.471044
Square hectometers	Ares	100
Square hectometers	Centares	10,000
Square hectometers	Hectares	1
Square hectometers	Sq cm	1×10^{8}
Square hectometers	Sq feet	107,638.7
Square hectometers	Sq inches	15,499,969
Square hectometers	Sq kilometers	0.01
Square hectometers	Sq meters	10,000
Square hectometers	Sq miles	0.003861006
Square hectometers	Sq rods	395.367
Square hectometers	Sq yards	11,959.85
Square inches	Ares	6.4516258×10^{-6}
Square inches	Hectares	6.4516258×10^{-8}
Square inches	Sq cm	6.4516258
Square inches	Sq decimeters	0.064516258
Square inches	Sq kilometers	$6.4516258 \times 10^{-10}$
Square inches	Sq meters	0.00064516258
Square inches	Sq mm	645.16258

TO CONVERT FROM	TO	MULTIPLY BY
Sq kilometers	Acres	247.1044
Sq kilometers	Ares	10,000
Sq kilometers	Hectares	100
Sq kilometers	Sq cm	1×10^{10}
Sq kilometers	Sq feet	1.076387×10^{7}
Sq kilometers	Sq inches	1.5499969×10^{9}
Sq kilometers	Sq meters	1,000,000
Sq kilometers	Sq miles	0.3861006
Sq kilometers	Sq mm	1×10^{12}
Sq kilometers	Sq rods	39,536.7
Sq kilometers	Sq yards	1.195985×10^{6}
Sq meters	Acres	0.0002471044
Sq meters	Ares	0.01
Sq meters	Centares	1
Sq meters	Hectares	0.0001
Sq meters	Sq cm	10,000
Sq meters	Sq decameters	0.01
Sq meters	Sq feet	10.76387
Sq meters	Sq inches	1549.9969
Sq meters	Sq kilometers	0.000001
Sq meters	Sq miles	3.861006×10^{-7}
Sq meters	Sq mm	1,000,000
Sq meters	Sq rods	0.0395367
Sq meters	Sq yards	1.195985
Square miles	Ares	25,899.98
Square miles	Hectares	258.9998
Square miles	Sq cm	2.589998×10^{10}
Square miles	Sq decimeters	2.589998×10^{8}
Square miles	Sq kilometers	2.589998
Square miles	Sq meters	2,589,998
Square miles	Sq mm	2.589998×10^{12}
Square millimeters	Circ mils	1973.52
Square millimeters	Sq cm	0.01
Square millimeters	Sq feet	1.076387×10^{-5}

TO CONVERT FROM	TO	MULTIPLY BY
Square millimeters	Sq inches	0.0015499969
Square millimeters	Sq kilometers	1×10^{-12}
Square millimeters	Sq meters	0.000001
Square millimeters	Sq miles	3.861006×10^{-13}
Square millimeters	Sq mils	1549.9969
Square millimeters	Sq rods	3.95367×10^{-8}
Square millimeters	Sq yards	1.195985×10^{-6}
Square mils	Sq cm	6.4516258×10^{-6}
Square mils	Sq mm	0.00064516258
Square rods	Ares	0.2529295
Square rods	Hectares	0.002529295
Square rods	Sq cm	2.529295×10^{5}
Square rods	Sq kilometers	2.529295×10^{-5}
Square rods	Sq meters	25.29295
Square yards	Ares	0.008361307
Square yards	Hectares	8.361307×10^{-5}
Square yards	Sq cm	8361.307
Square yards	Sq decameters	0.008361307
Square yards	Sq kilometers	8.361307×10^{-7}
Square yards	Sq meters	0.8361307
Square yards	Sq mm	836,130.7
Steres	Cu meters	1
Steres	Decasteres	0.1
Steres	Decisteres	10
Steres	Liters	999.973
Tons (long)	Grams	1.01604704×10^{6}
Tons (long)	Kilograms	1016.04704
Tons (long)	Tons (metric)	1.01604704
Tons (metric)	Grams	1,000,000
Tons (metric)	Kilograms	1000

TO CONVERT FROM	TO	MULTIPLY BY
Tons (metric)	Pounds (apoth or troy)	2679.229
Tons (metric)	Pounds (avdp)	2204.62234
Tons (metric)	Tons (long)	0.98420640
Tons (metric)	Tons (short)	1.1023112
Tons (short)	Grams	9.0718486×10^5
Tons (short)	Kilograms	907.18486
Tons (short)	Tons (metric)	0.90718486
Wine gallons	Liters	3.785332
Yards	Centimeters	91.44018
Yards	Hectometers	0.009144018
Yards	Kilometers	0.00091440183
Yards	Meters	0.9144018
Yards	Microns	914,401.8
Yards	Millimeters	914.4018
Yards	Millimicrons	9.114018×10^8

Abbreviations Used
in This Book

A—ampere (s)
a—are (s)
a—atto-
abs—absolute
apoth—apothecaries'
  weight
atm—atmosphere
avdp—avoirdupois weight
Btu—British thermal unit
C—Celsius (formerly
  centigrade)
c—centi-
ca—centare (s)
cal—calorie (s)
cc—cubic centimeter (s)
cd—candela (s)
cg—centigram (s)
circ—circular
cl—centiliter (s)
cm—centimeter (s)

cm^2—square centimeter (s)
cm^3—cubic centimeter (s)
cu—cubic
cu cm—cubic centimeter (s)
cu dkm—cubic
  decameter (s)
cu dm—cubic decimeter (s)
cu ft—cubic foot, cubic feet
cu hm—cubic
  hectometer (s)
cu km—cubic kilometer (s)
cu m—cubic meter (s)
cu mm—cubic
  millimeter (s)
cu yd—cubic yard (s)
d—deci-
dg—decigram (s)
dk—deca-
dkg—decagram (s)
dkl—decaliter (s)

dkm—decameter (s)
dks—decastere (s)
dl—deciliter (s)
dm—decimeter (s)
ds—decistere (s)
F—Fahrenheit
f—femto-
ft—foot, feet
G—giga-
g—gram (s)
gal—gallon (s)
gm—gram (s)
h—hecto-
ha—hectare (s)
hg—hectogram (s)
hl—hectoliter (s)
hm—hectometer (s)
hp—horsepower
hr—hour (s)
i.e.—that is
in.—inch (es)
IT—International Steam
  Table
K—kelvin (s)
k—kilo-
kg—kilogram (s)
kl—kiloliter (s)
km—kilometer (s)
km^2—square kilometer (s)
km^3—cubic kilometer (s)
kw-hr—Kilowatt-hour (s)
l—liter (s)
liq—liquid
M—mega-
m—meter (s)
m—milli-
m^2—square meter (s)

m^3—cubic meter (s)
mg—milligram (s)
min—minute (s)
ml—milliliter (s)
Mm—megameter (s)
mm—millimeter (s)
MT—metric ton (s)
my—myria-
mym—myriameter (s)
μ—micro-
μ—micron (s)
μg—microgram (s)
μm—micrometer (s)
N—newton (s)
n—nano-
p—pico-
q—quintal (s)
s—second (s)
s—stere (s)
SI—Système International
sq—square
sq cm—square
  centimeter (s)
sq dkm—square
  decameter (s)
sq ft—square foot,
  square feet
sq in.—square inch (es)
sq km—square kilometer (s)
sq m—square meter (s)
sq mm—square
  millimeter (s)
T—tera-
t—metric ton (s)
U.S.—United States
yd—yard (s)

Index